Dr TF Walsh
Charles Clifford Dental Hospital
Wellesley Road
SHEFFIELD S10 2SZ
Tel: 0742 670444
Fax: 0742 665326

IRL PRESS
—at—
OXFORD UNIVERSITY PRESS
Oxford New York Tokyo

RECENT ADVANCES IN THE STUDY OF DENTAL CALCULUS

Proceedings of a Workshop
6 – 9 November, 1988
Noordwijkerhout, The Netherlands

Edited by
J.M.ten Cate

Organized by
THE RESEARCH GROUP ON
SURFACE AND COLLOID PHENOMENA
IN THE ORAL CAVITY
under the auspices
of THE COUNCIL OF EUROPE

IRL Press
Eynsham
Oxford
England

British Library Cataloguing in Publication Data
Recent advances in the study of dental calculus.
1. Man. Teeth. Calcium phosphates
I. Cate, J. M. ten   II. Research Group on Surface and Colloid
Phenomena in the Oral Cavity
612'.311

ISBN 0 19 963122 0

The correct manner by which to refer to a paper from this
publication is: Author of paper, Title of paper, 'Recent Advances
in the Study of Dental Calculus', IRL Press (Oxford), page
numbers, 1989.

Printed by Information Press Ltd, Oxford, England.

CONTENTS

ACKNOWLEDGEMENTS

The 1988 Workshop of the Research Group on Surface and Colloid Phenomena in the Oral Cavity, part of the Study Group on Surface Chemistry and Colloids, Committee of Science Technology, Parliamentary Assembly of the Council of Europe, was held at the Leeuwenhorst Congres Center, Noordwijkerhout, near Leiden, The Netherlands, November 6 – 9, 1988, under the chairmanship of Professor Bob (J.M.) ten Cate, Academic Centre for Dentistry Amsterdam, Amsterdam, The Netherlands.

This Workshop was financially supported by the Council of Europe; Colgate-Palmolive Company (USA); Gaba International Ltd (Switzerland); Henkel KGaA (Germany); Intradal N.V. (The Netherlands); Johnson & Johnson Dental Care (USA); The Coca-Cola Company (USA); The Procter & Gamble Company (USA); Unilever Research (UK); Warner Lambert Company (USA); who are all most gratefully acknowledged.

Thanks are also due to Mrs Anneke Reijnders and other members of the Department of Cariology & Endodontology, Academic Centre for Dentistry Amsterdam, for their contribution to the organization of the Workshop.

PREFACE

The Research Group on Surface and Colloid Phenomena in the Oral Cavity met on November 6 – 9, 1988 at Noordwijkerhout, near Leiden, The Netherlands. This Group was established in 1975 under the auspices of the Parliamentary Assembly of the Council of Europe Committee on Science and Technology. Its objectives are to create contacts between surface and colloid scientists and dental researchers in order to promote a multidisciplinary approach and the cross-fertilization of ideas and expertise relevant to the oral cavity and so help to put the prevention of dental caries and periodontal disease on a scientific basis.

The Group has met on 16 occasions since its inception in 1975, in France, Denmark, Greece, The Netherlands, Norway, Sweden, Turkey and the United Kingdom. These are the eleventh proceedings published by the Group and the ninth to be published annually in book form by IRL Press since 1980.

This publication is the outcome of the meeting at Noordwijkerhout, which was organized as a Workshop that focused specifically on dental calculus. In recent years the prevention of calculus has received considerable commercial attention. Many brands of 'anti-tartar' toothpastes are now being sold worldwide. Clinical studies have shown that the active ingredients of such pastes are very effective in inhibiting calculus formation. At the same time concern has arisen that these additives, being crystallization inhibitors, may hamper the beneficial effects of toothpaste-derived fluoride in the prevention of caries. The renewed interest in dental calculus has initiated new projects on its etiology. Presently new epidemiological as well as (bio)chemical methodologies are available to study the formation of calculus and to determine an association between physiological parameters and the occurrence of dental calculus in a population.

The various topics mentioned above were discussed during five sessions entitled:
I. Mechanism of calculus formation; II. Calculus in relation to plaque and bacteria; III. Composition and formation of calculus; IV. Calculus inhibition; and V. Clinical aspects of calculus (vs caries).

The editor wishes to thank the authors for their contributions, as well as the participants in the 1988 Workshop for their contributions to the constructive and animated discussions that took place.

J.M.ten Cate

Chairman
Academic Center for Dentistry Amsterdam,
Amsterdam, The Netherlands

J.Arends	University of Groningen, Laboratory for Materia Technica, Antonius Deusinglaan 1, 9713 AV Groningen, The Netherlands
V.Bieri	Gaba International Ltd, Grabetsmattweg, CH-4106 Therwil, Switzerland
W.D.Bowman	The Procter & Gamble Company, Sharon Woods Technical Center, 11511 Reed Hartman Highway, Cincinnati, Ohio 45241, USA
B.D.Boyan	The University of Texas Health Science Center, 7703 Floyd Curl Drive, San Antonio, Texas 78284, USA
A.Burger	Lever Research & Development Center, 45 River Road, Edgewater, New Jersey 07020, USA
T.van de Burgt	Intradal N.V., Brabantsestraat 17, 3812 PJ Amersfoort, The Netherlands
H.J.Busscher	University of Groningen, Laboratory for Materia Technica, Antonius Deusinglaan 1, 9713 AV Groningen, The Netherlands
J.M.ten Cate	Academic Centre for Dentistry Amsterdam, Department of Cariology & Endodontology, Louwesweg 1, 1066 EA Amsterdam, The Netherlands
J.J.M.Damen	Academic Centre for Dentistry Amsterdam, Department of Cariology & Endodontology, Louwesweg 1, 1066 EA Amsterdam, The Netherlands
F.C.M.Driessens	Catholic University Nijmegen, Department of Oral Biomaterials, PO Box 9101, 6500 HB Nijmegen, The Netherlands
R.Duckworth	Unilever Research, Port Sunlight Laboratory, Quarry Road East, Bebington Wirral, Merseyside L63 3JW, UK
W.M.Edgar	University of Liverpool, Department of Dental Sciences, PO Box 147, Liverpool L69 3BX, UK
S.Edwardsson	University of Lund, School of Dentistry, Department of Oral Microbiology, S-214 21 Malmö, Sweden

G.Embery	University of Wales, Department of Basic Dental Science, Heath Park, Cardiff CF4 4XY, UK
J.D.B.Featherstone	Eastman Dental Center, 625 Elmwood Avenue, Rochester, NY 14620, USA
O.Fejerskov	The Royal Dental College, Department of Dental Pathology & Operative Dentistry, Vennelyst Boulevard, DK-8000 Aarhus C, Denmark
D.Gaare	University of Oslo, Faculty of Dentistry, Geitmyrsveien 71, 0455 Oslo 4, Norway
A.Gaffar	Colgate-Palmolive Research Center, 909 River Road, Piscataway, NJ 08854 – 5596, USA
B.R.Heywood	School of Chemistry, University of Bath, Claverton Down, Bath BA2 7AY, UK
G.S.Ingram	Unilever Research, Port Sunlight Laboratory, Quarry Road East, Bebington Wirral, Merseyside L63 3JW, UK
G.Koch	The Institute for Postgraduate Dental Education, Box 1030, S-551 11 Jönköping, Sweden
S.Mason	Colgate-Palmolive Company, 909 River Road, Piscataway, NJ 08854, USA
E.C.Moreno	Forsyth Dental Center, 140 Fenway, Boston, MA 02115, USA
M.Morita	Okayama University Dental School, Department of Preventive Dentistry, Shikato-cho, Okayama 700, Japan
G.Payonk	Johnson & Johnson Dental Care Company, Route #1 South, North Brunswick, New Jersey 08902, USA
J.P.Piessens	ACP bv, PO Box 16299, 2500 AE The Hague, The Netherlands
W.Plöger	Henkel KGaA, Postfach 1100, D-4000 Düsseldorf 1, Federal Republic of Germany
D.Purdell-Lewis	Unilever Research, Redstone, 16 Tower Road North, Heswall Wirral, Merseyside L63 3JW, UK
G.Rølla	University of Oslo, Faculty of Dentistry, Geitmyrsveien 71, 0455 Oslo 4, Norway
A.A.Scheie	University of Oslo, Department of Microbiology, PO Box 1052, Blindern, 0316 Oslo 3, Norway

G.K.Stookey   Indiana University School of Dentistry, Oral Health
                          Research Institute, 415 Lansing Street,
                          Indianapolis, Indiana 46202, USA

L.D.Swain   The University of Texas Health Sciences Center,
                          7703 Floyd Curl Drive, San Antonio, Texas
                          78284, USA

A.Tatevossian   18 Thornhill Road, Llanishen, Cardiff CF4 6PF, UK

A.Thylstrup   Royal Dental College Copenhagen, Department of
                          Cariology & Endodontology, 20 Nørre Allé,
                          DK-2200 Copenhagen N., Denmark

H.M.Uyen   University of Groningen, Laboratory for Materia
                          Technica, Antonius Deusinglaan 1, 9713 AV
                          Groningen, The Netherlands

T.Watanabe   Okayama University Dental School, Department of
                          Preventive Dentistry, Shikata-cho, Okayama, 700
                          Japan

D.J.White   The Procter & Gamble Company, Sharon Woods
                          Technical Center, 11511 Reed Hardman Highway,
                          Cincinnati, Ohio 45241, USA

M.A.O.W.Zoon   Academic Centre for Dentistry Amsterdam,
                          Department of Cariology & Endodontology,
                          Louwesweg 1, 1066 EA Amsterdam, The
                          Netherlands

J.M. ten Cate

Research on dental calculus: why?

Department of Cariology and Endodontology,
Academic Centre for Dentistry Amsterdam,
Amsterdam, The Netherlands

Abstract

After a period in which few scientists were involved in calculus
this field of research is now showing a remarkable increase in
activity. Possibly this is the spin-off of a commercial interest
on this topic. However, several other reasons can be given to explain this phenomenon. Some of these are discussed briefly in
this introduction.

The importance of calculus removal to maintain or to improve the
periodontal health validates studies on the etiology and prevention of calculus, as well as a search for possible indicators
for calculus formation.

The dynamics of de- and remineralisation at the enamel-oral
fluid-interface are presently changing, as reflected by a decreasing prevalence of caries. In principal this could also result in changing patterns and prevalence of dental calculus.

If an effective therapy to prevent calculus could be formulated
this would substantially diminish the expenses involved in calculus removal by the dental profession. Lastly, calculus, being a
mineralisation process in which bacteria are involved, may be of
interest scientifically, as it serves as a model for other forms
of bio- or pathological mineralisation.

Dental calculus has been an important area of dental research for many years. The original rationale for studying the mechanisms of dental calculus formation and for trying to find ways for its prevention was its assumed correlation with periodontal diseases. Epidemiological studies which were aimed at finding such a relationship lead to the conclusion that not calculus but plaque was the cause of gingivitis and more advanced periodontal disease. Because of this finding and the increased caries prevalence in the 1960's the interest in calculus became dwindling, until in the 1980's the topic of dental calculus was revitalised.

It is interesting to examine the reasons behind this changed attitude. Why has calculus suddenly attracted the attention of the dental and bio-medical researchers? Were the increased research activities primarily generated by a commercial interest in 'anti-tartar' products?

In my opinion at least four other reasons can be given:

- A revaluation of the possible relationship between calculus and the initiation and progression of periodontal disease:
A recent paper on dental calculus (1) reviews the data on this topic, which has appeared since Schroeder's 1969 textbook 'Formation and Inhibition of Dental Calculus'(2).
Throughout this period epidemiological studies have been undertaken to determine a possible correlation between plaque index, calculus index and periodontal disease. In spite of the inconsistencies in the scoring methods of each of these parameters the general conclusion is that these correlations exist. However, it is generally emphasised that this does not necessarily imply a direct causal relationship. Subgingival calculus is not considered the cause of gingivitis, but is more likely the result of the combination of plaque and the changed local environment due to gingival inflammation. Supragingival calculus forms through the mineralisation of dental plaque even when there are no signs of periodontal disease.
Although calculus may not be the cause of periodontal disease, it could well be a factor in its progression or persistence. Different authors found that a removal of calculus was associated with an improvement of periodontal health, whereas the removal of plaque alone was far less effective. Likely this is due to the fact that plaque cannot be removed satisfactorily when there is a calculus deposit. A possible second explanation for this finding could be the microscopic observation of the porosity of calculus. Mandel and Gaffar (1) describe calculus as 'a toxic waste dump site and in a sense a slow release device delivering pathogenic products'.
Whatever the relationship between calculus and periodontal disease it should be stressed that the presence of calculus hampers the mechanical removal of plaque and thereby interferes with the prevention and healing of periodontal infections.

- Calculus might serve as a model for pathological or bacteria associated calcification:
Many aspects of the mineralisation of plaque and the possible active role of bacteria, bacterial components or, in general, high molecular weight components, in calculus formation have a

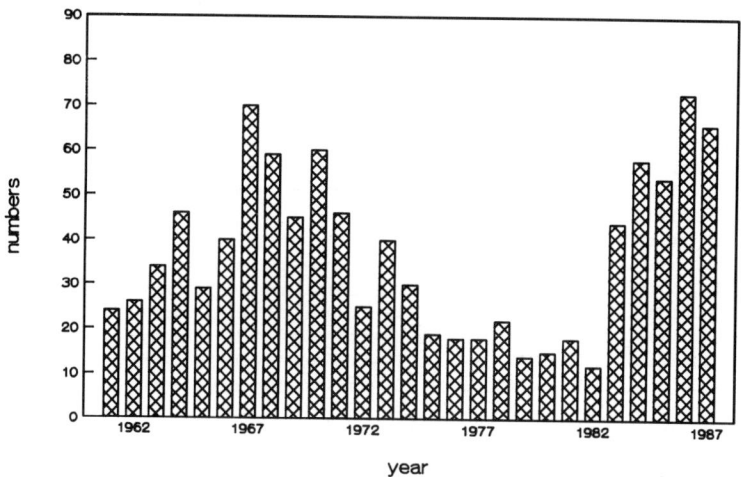

Figure 1: Number of articles included in the annual volumes of the Index Dental Literature, categorised by the keyword 'Dental Calculus'.

similarity in biomineralisation and in the formation of other calcium containing deposits in the body. Among these are the formation of renal stones, arthritis and aorta calcification. With respect to renal stone formation it has been shown that glycosaminoglycans and mucoproteins interfere with calcium oxalate crystal growth and stimulate the aggregation of small crystals, thereby enhancing the formation of renal stones (3). The regulatory activity of components of the bacterial cell membrane have an analogue in the deposition of the mineral phase of calcified tissues.
Although a bacterial involvement in the formation of dental calculus has been demonstrated it still remains to be answered whether the precipitation of calcium phosphates in mineralising dental plaque is primarily determined by physico-chemical parameters or that it is directed by processes actively induced by bacteria.

- The changing caries prevalence may contribute to calculus formation:
With the introduction of a wide range of fluoride topicals it has become possible to prevent dental caries, without too much emphasis on improving oral hygiene. Although the overall sugar intake is not reduced there are indications that the number of cariogenic in-between-meal-snacks is reduced. Thus a situation arises where the amounts of plaque are not necessarily reduced, although the acid-base and the dissolution-reprecipitation balance are shifted. An important factor in this respect is fluoride, which contributes to the formation of less soluble calcium phosphate phases, in the dental hard tissues and possibly in the plaque. Considering that the precipitation of calcium

phosphates generally proceeds through a series of minerals with decreasing solubility it is feasible that fluoride will cause the plaque deposits to be mineralised faster and in a less soluble form at an early stage.

- The development of a chemical method of calculus prevention could eliminate the necessity for professional calculus removal: This economic reason for studying calculus and its prevention can be illustrated by figures of the costs of professional removal of calculus. As an example the data of the Dutch 'sickfund' have been analysed (4). This is a subsidised national health system for low and medium income families in The Netherlands, covering about two thirds of the population. The system gives refunds to dentists for the majority of the dental treatments. The information pertaining to calculus removal show that 50% of the patients are treated for calculus removal, amounting to 10% of the total dental budget (i.e. 35 million US dollars). The data show an increase by about 8% over the last 3 years, compared to an average overall decrease of the dental budget by 1%. One could argue that these figures are not a true estimate of the severity of the 'disease'. With the decreasing number of caries treatments the dentists are likely to shift their attention to calculus removal. This, however, implies an unlikely previous neglect of a possible health hazard. Critics may also argue that calculus removal is an activity that is very difficult to verify, which may therefore presently be claimed more frequently than absolutely needed.

Whatever the reason, the increased interest in dental calculus is very apparent as illustrated by the numbers of scientific publications for which calculus was indicated as a keyword (Figure 1). Without doubt, so much new information regarding calculus, or concerning biological mineralisation with a spin-off to calculus, has been generated that a publication on this topic updating the available knowledge is justified.

References:
1. Mandel,I.D. and Gaffar, A.: Calculus revisited, a review. J. Clin. Periodontolgy 13:249-257 (1986).

2. Schroeder,H.E.: Formation and inhibiton of dental calculus, Hans Huber Publishers, Berne, 1969.

3. Robertson,W.G., Scurr,D.S. and Bridge,C.M.: Factors inflencing the crystallization of calcium oxalate in urine- critique. J.Cryst.Growth 53:182-194 (1981).

4. Jaarverslag Nederlande Ziekenfondsraad 1987.

SESSION I.
Mechanism of Calculus
Formation

F.C.M.Driessens
R.M.H.Verbeeck*

Possible pathways of mineralization of dental plaque

Department of Oral Biomaterials, Catholic University,
PO Box 9101, 6500 HB Nijmegen, The Netherlands
and Laboratory for Analytical Chemistry,
State University, Ghent, Belgium

ABSTRACT

Previous studies have shown that mineralization in bone,
dentin and internal pathological calcifications starts with
the deposition of octocalcium phosphate OCP which later on
transforms into a mixture of other calcium phosphates like
(amorphous) magnesium whitlockite and some carbonated apatites. However, in tooth enamel mineralization starts probably
with the deposition of brushite or dicalcium phosphate dihydrate DCPD onto which later on very quickly is deposited a
mixture of magnesium whitlockite, carbonated apatites and
calcium-deficient apatites.
Apparently, both brushite and OCP occur in human dental
calculus and, hence, both mechanisms as mentioned for the calcified tissues, may occur during the mineralization of dental
calculus. The kinetics is such that upon local supersaturation
brushite and OCP can precipitate very quickly and that the
other calcium phosphates use them as precursor crystals so
that further mineralization takes place either by ongrowth of
the other less soluble phases (brushite in tooth enamel) or by
topotactical transformation into the other less soluble phases
(OCP in bone, dentin and pathological calcifications). This
view is corroborated by comparison of in vitro solubility data
for the different calcium phosphates with the literature data
of plaque fluid composition and pH which show, under which
conditions plaque fluid can become supersaturated with the
calcium phosphates in question, while simultaneously the
plaque bacteria break down calcification inhibitors (e.g. by
pyrophosphatase activity) and organic components of the plaque
serve as templates for nucleation of these minerals (e.g. especially some phospholipids). When salivary pH is above 8 as
occurs in some animals, the preferent mineral in dental calculus is calcite. This is explained by the form of the ion activity product diagram for calcium phosphates and carbonates.
It indicates that a local increase of pH whereby the degree of
supersaturation is increased is coresponsible for plaque mineralization and calculus formation.

* Research Associate NFSR (Belgium)

Recent Advances in the Study of
Dental Calculus

STRUCTURE AND COMPOSITION OF THE MINERAL IN HUMAN DENTAL CALCULUS

Table I shows the prevalence of mineral structures occurring in human dental calculus according to quantitative X-ray diffraction studies reported in the literature. There is some difference between subgingival and supragingival plaque: the former contains more whitlockite and the latter more apatite and more frequently brushite[7]. As far as the morphology is concerned, whitlockite occurs as bulk crystals and octocalcium phosphate OCP as platelet-shaped crystals, often epitactical to platelet-shaped apatite. In addition needle-shaped crystals of brushite and of apatite occur[4,7]. The volume percentage of mineral in supragingival calculus can vary from 16 to 80%, that of subgingival calculus from 46 to 83%[8].

Table II contains data about the inorganic composition of human dental calculus, as mentioned in the literature. A significant positive correlation has been found between magnesium content and whitlockite abundance and a negative correlation between magnesium and OCP. The carbonate content has been found to be positively correlated with apatite abundance and negatively with OCP[12]. X-ray diffraction[13] and electron microprobe analysis[7] showed that magnesium occurred only in the whitlockite phase. According to LeGeros[14] the magnesium content of human dental calculus is the result of partial substitution of magnesium for calcium in whitlockite, whereas the carbonate content is the result of substitution in apatite. Taking into consideration that supragingival calculus contains more apatite[7] and more sodium[10,11] than subgingival calculus, one can conclude that the sodium content of dental calculus is probably also the result of substitution of sodium for calcium in apatite. These data enable a closer analysis of the phase composition of the mineral in dental calculus, especially when this is compared to the mineral of tooth enamel, bone, dentin, salivary calculi and other pathological calcifications[15,16].

As far as the age of the calcifying plaque is concerned, brushite and OCP have a higher prevalence and content in younger samples of dental calculus. Brushite vanishes rapidly with increasing age, whereas OCP occurs still in old samples of dental calculus[4,17]. For these reasons it is thought[18] that

Table I. PREVALENCE OF MINERAL STRUCTURES IN HUMAN DENTAL CALCULUS

Apatite	Whitlockite	OCP	Brushite	Reference
100	80	-	14	Tovborg Jensen & Danø[1]
100	65	-	-	Forsberg et al.[2]
99.5	80.7	94.8	43.6	Rowles[3]
98	71	95	44	Schröder and Baumbauer[4]
93	86	95	18	Grøn et al.[5]
100	-	-	40	Ölzner et al.[6]

Table II. INORGANIC COMPOSITION OF HUMAN DENTAL CALCULUS
(% OF DRY WEIGHT)

Ca	P	Mg	CO_3	Na	References
29-32	16-17.5	0.9-1.1	2	n.d.	Glock & Murray[9]
18-29	13-17	n.d.	n.d.	1.9-2.6	Little et al.[10a]
25-31	14-17	n.d.	n.d.	1.2-1.8	Little & Hazen[11b]
25-29	13-16	0.4-0.9	1-3	n.d.	Grøn et al.[5a]
23-28	13-16	0.5-1.3	1.5-2.5	n.d.	Grøn et al.[5b]

a   supragingival
b   subgingival
n.d. not determined

OCP and brushite are formed as precursor minerals during the
initial stage of calcification of dental plaque and that they
are gradually hydrolyzed and transformed into apatite and/or
whitlockite.

COMPARISON WITH THE MINERAL IN CALCIFIED TISSUES AND OTHER PA-
THOLOGICAL CALCIFICATIONS

According to thermodynamical considerations about the free
energy of solid solutions[19-21], it is highly unlikely that
solid solutions are formed by slow precipitation of minerals
from complex aqueous solutions like the body fluids. One
should rather expect that a mixture of minerals is formed,
each being close to a certain stoichiometry[22-24]. Such minerals should be predominantly end members of certain series of
solid solutions. In this way one arrives at the phases mentioned in Table III as being serious candidates for the occurrence in biominerals.
Several of the phases of Table III can form epitactical or
topotactical mixtures, like calcite with dolomite or brushite
with any of the apatites[25] or OCP with any of the apatites[26]
or the apatites among each other. Therefore, many mineral particles in calcified tissues and pathological calcifications
may contain different domains with different stoichiometries.
Taking further into account that the particle size is mostly
so small that these separate domains cannot be detected by
physical methods, one is stuck with the situation that only by
chemical methods more insight into the matter of biominerals
can be gained. Chemical methods which are feasible in this
case are
(a) studies of systematic variations in composition of the
    biominerals
(b) studies on the solubility behaviour of the biominerals.
We have applied a number of studies to reveal systematic
variations in the composition of biominerals of tooth
enamel[27-31], dentin[27,28], bone[32,33], aorta and heart valve
calcifications[16] and salivary calculi[34]. It appeared that all
variations in composition could be explained by assuming that

Table III.   CANDIDATE PHASES FOR OCCURRENCE IN BIOMINERALS

Abbre- viation	Formula	Name
-	$CaCO_3$	calcite or aragonite
DOL	$CaMg(CO_3)_2$	dolomite
DCPD	$CaHPO_4-2H_2O$	brushite
OCP	$Ca_8(HPO_4)_2(PO_4)_4.5H_2O$	octocalcium phosphate
OHA	$Ca_{10}(PO_4)_6(OH)_2$	hydroxyapatite
DOHA	$Ca_9(HPO_4)(PO_4)_5(OH)$	defective hydroxyapatite
HCDOHA	$Ca_9(PO_4)_{4.5}(CO_3)_{1.5}(OH)_{1.5}$	heavily carbonat.def.hydr.
NCCA	$Ca_{8.5}Na_{1.5}(PO_4)_{4.5}(CO_3)_{2.5}$	Na- and CO_3-cont. apatite
MWH	$Ca_9Mg(HPO_4)(PO_4)_6$	magnesium whitlockite

these biominerals consisted of a limited number of the candidate phases of Table III. The results are compiled in Table
IV. Two groups of biominerals can be distinguished: one with
OCP as the probable precursor[33-36], the other with DCPD as the
probable precursor[34,35,37]. On the basis of these data it is
hypothesized here that the mineral of human dental calculus
can contain the whole scala of phases mentioned in Table IV,
whereby the spatial distribution depends on the original spatial distribution of the precursors DCPD and OCP, like the
situation occurring in salivary calculi[34].

SOLUBILITY BEHAVIOUR OF BIOMINERALS AND MECHANISM OF FORMATION
OF DENTAL CALCULUS

    The solubility behaviour of biominerals and of their syn-
thetic analogs has been studied more or less systematically by
many investigators during the past 60 years[15,16]. Visualiza-
tion of the relative solubility of calcium phosphates is done
most easily in a so-called activity product diagram. In such a
diagram the logarithm of the ion activity product for hydroxyapatite
$logI_p^{OHA}$ = 10 log a(Ca^{2+}) + 6 log a (PO_4^{3-}) + 2 log a (OH$^-$)
in which a(X) represents the activity of ion X in solution, is

Table IV.   PROBABLE PHASE COMPOSITION OF SOME HUMAN BIOMINER-
ALS

	NCCA	MWH	HCDOHA	OCP	DCPD	OHA	DOHA	DOL
Bone	+	+	+	+				
Dentin	+	+	+					
Aorta and heart valve	+	+	+					
Salivary calculi(partly)	+	+	+	+				
Salivary calculi(partly)	+				+		+	+
Tooth enamel	+				+	+	+	+

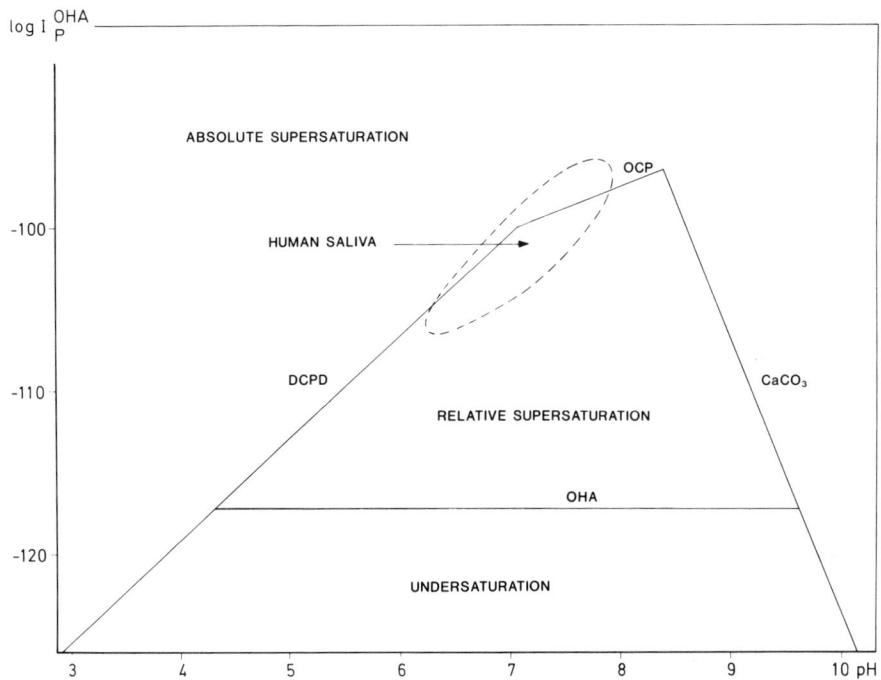

Figure 1. Location of human saliva in the activity product diagram for calcium phosphates and carbonates.

plotted versus pH. A plot of the biologically relevant region in terms of such a diagram is given in Fig. 1. This region is bordered on the side of absolute supersaturation by the solubility curves for DCPD, OCP and $CaCO_3$ which are fast nucleators. If a body fluid passes those lines, precipitation of mineral becomes inevitable, unless inhibitors are present which prevent it up to a higher level of supersaturation. Under physiological conditions the location of most body fluids is within the field of relative supersaturation[36,38]. An important exception is formed by human saliva, especially when excretion is stimulated[39,40]. Not only supersaturation with OCP but also with brushite can occur in human saliva. Some of the precipitation inhibitors have been identified[39].

The solubility behaviour of the most important biominerals is indicated in Fig. 2. That of living bone equals that of finely dispersed OCP[33,41], whereas that of dentin mineral[42] is most representative for HCDOHA which predominates in that mineral. The solubility behaviour of tooth enamel mineral is rather complex. Upon contact with acidic solutions, the Na and the Mg containing part of the mineral dissolve first[43-47], possibly due to the fact that these parts contain carbonate. After that stage the solubility of powdered tooth enamel is

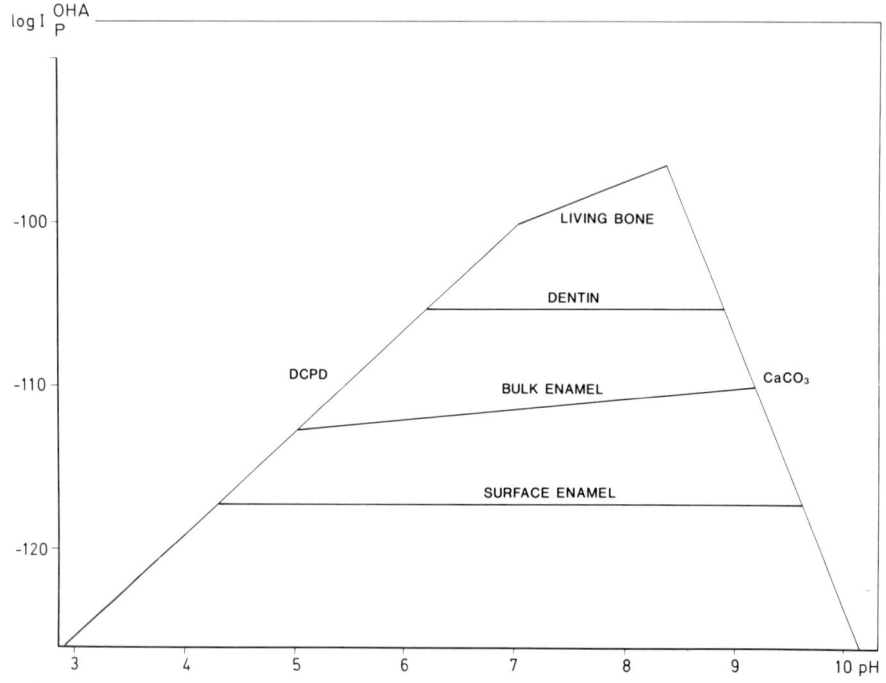

Figure 2.   Solubility behaviour of some biominerals.

controlled by DOHA[41,28]. However, whole human tooth enamel dissolves as if it were pure OHA[49], which is in accordance with the fact that the surface layer contains the OHA phase of enamel mineral[29]. Therefore, a wide variation of solubilities is found for tooth enamel mineral, ranging from a value of -105 to one of -118 for $\log I_p^{OHA}$[50].

The ranges of solubility found for biominerals are in line with expectations based on their phase composition (Table IV). For most of them the formation product as well as the solubility product are known at the moment, except for some carbonated species like NCCA and HCDOHA. Formation product and solubility product have been described for OCP, DCPD, OHA, DOHA, MWH, DOL and $CaCO_3$[15,16,30,51]. In fact, these data present the second piece of hard chemical evidence for the occurrence of these phases in biominerals mentioned in Table IV and thus for their probable occurrence in dental calculus or e.g. carious dentin[59].

What then is the mechanism of formation of dental calculus as far as the physicochemical factors are concerned? Most investigators have no doubt that the main source of the inorganic ions forming the mineral of dental calculus is the saliva. Older theories were in favour of the idea that locally the plaque could have a pH higher than that of saliva, either by

Table V.   SALIVARY pH (UNSTIMULATED) AND PHASE COMPOSITION OF DENTAL CALCULUS IN MAMMALS[58]

Mammal	Salivary pH	Phase composition of dental calculi
Miniature pig	8	$CaCO_3$
Domestic pig	8	$CaCO_3$
Dog	8	$CaCO_3$ (+ NCCA)
Primates	7.5	NCCA (+ MWH)
Sheep	7.5	NCCA + MWH (+ DCPD)
Man	6.8	NCCA + MWH (+ OCP + DCPD)

CO_2 loss or by ammonia formation[52]. Such an increase of pH will automatically result in precipitation of phosphates and carbonates. In later theories it was thought that plaque could contain certain organic molecules functioning as templates for the precipitation of minerals[53]. A third possibility is, of course, that the plaque is able to break down nucleation inhibitors.

There is some proof in the literature for the latter two proposed mechanisms, as will be dealt with by other authors. However, there is ample evidence that a local increase in pH is primary responsible for dental calculus formation. Kleinberg and Jenkins[54] investigated the pH of resting plaque. In 85% of the cases it was higher than that of the surrounding saliva by 0.5 to 1 unit. This is probably due to proteolytic activity in which urea, ammonia and amines are formed[55]. This view is in line with the findings of Watanabe et al.[56] according to which protease activity in saliva is positively correlated with calculus index. In the front teeth mouth breathing and loss of CO_2 is an important factor.

Another piece of evidence for the local increase of pH as primary cause of calculus formation comes from a comparison of dental calculus between men and animals. LeGeros and Shannon[57] mentioned that alkalification of human saliva resulted in a precipitate which contained whitlockite and apatite. However, alkalification of dog saliva which has a pH of 8 or more results in the formation of $CaCO_3$ as should be expected on the basis of Fig. 1. Dental calculus formed in vivo in a number of species has been collected by Driessens et al.[58]. That study proved that calcite was the predominant structure in dental calculus, when the salivary pH was 8 or higher, whereas a mixture of apatite and whitlockite with small amounts of OCP and brushite were found, when the salivary pH was lower than 8. See Table V.

REFERENCES

1. Tovborg Jensen, A. and Danø, M. (1954) Crystallography of dental calculus and the precipitation of certain calcium phosphates. J. dent. Res. 33, 741-750.
2. Forsberg, A., Lagergren, C. and Lonnerblad, T. (1960) Dental calculus. Oral Surg. 13, 1051-1060.
3. Rowles, S.L. (1964) Biophysical studies on dental calculus

in relation to periodontal disease. Dent. Pract. dent. Rec. 15, 2-7.

4. Schröder, H.E. and Baumbauer, H.U. (1966) Stages of calcium phosphate crystallisation during calculus formation. Archs. oral Biol. 11, 1-14.
5. Grøn, P., Campen, G.J. van and Lindstrom, I. (1967) Human dental calculus. Archs. oral Biol. 12, 829-837.
6. Ölzner, W., Hesse, A., Tscharnke, J. and Schneider, H.J. (1973) Struktur und Aufbau menschlichen Zahnsteins. Dt. Stomatol. 23, 8-16.
7. Sundberg, M. and Friskopp, J. (1985) Crystallography of supragingival and subgingival human dental calculus. Scand. J. dent. Res. 93, 30-38.
8. Friskopp, J. and Isacsson, G. (1984) A quantitative microradiographic study of mineral content of supragingival and subgingival dental calculus. Scand. J. dent. Res. 92, 25-32.
9. Glock, G.E. and Murray, M.M. (1938) Chemical investigation of salivary calculus. J. dent. Res. 17, 257-264.
10. Little, M.F., Casciani, C.A. and Rowley, J. (1963) Dental calculus composition. I. Supragingival calculus. J. dent. Res. 42, 78-86
11. Little, M.F. and Hazen, S.P. (1964) Dental calculus composition. II. Subgingival calculus. J. dent. Res. 43, 645-651.
12. Grøn, P. and Campen, G.J. van (1967) Mineral composition of human dental calculus. Helv. odont. Acta 11, 71-74.
13. Knuuttila, M., Lappalainen, R. and Kontturi-Närhi, V. (1980) Effect of Zn and Mg on the formation of whitlockite in human subgingival calculus. Scand. J. dent. Res. 88, 513-516.
14. LeGeros, R.Z. (1974) Variations in the crystalline components of human dental calculus. I. Crystallographic and spectroscopic methods of analysis. J. dent. Res. 53, 45-50.
15. Driessens, F.C.M. (1982) Mineral aspects of dentistry. Karger, Basel.
16. Driessens, F.C.M. and Verbeeck, R.M.H. (1989) Biominerals, CRC Press, Boca Raton.
17. Kaufman, H.W. and Kleinberg, I. (1973) X-ray diffraction examination of calcium phosphate in dental plaque. Calcif. Tissue Res. 11, 97-104.
18. Kani, T., Kani, M., Moriwaki, Y. and Doi, Y. (1983) Microbeam X-ray diffraction analysis of dental calculus. J. dent. Res. 62, 92-95.
19. Driessens, F.C.M. (1968) Thermodynamics and defect chemistry of some oxide solid solutions. I. Nearest neighbour interactions and the effect of substitutional disorder. Ber. Bunsenges. Phys. Chem. 72, 754-764.
20. Driessens, F.C.M. (1968) Thermodynamics and defect chemistry of some oxide solid solutions. II. Pair interactions. Ber. Bunsenges. Phys. Chem. 72, 764-773.
21. Driessens, F.C.M. (1968) Thermodynamics and defect chemistry of some oxide solid solutions. III. Defect equilibria and the formalism of pair interaction. Ber. Bunsenges. Phys. Chem. 72, 1123-1133.

22. Driessens, F.C.M. and Verbeeck, R.M.H. (1984) Solubility behaviour of ionic solid solutions and their precipitation from aqueous solution. Bull. Soc. Chim. Belg. 93, 85-97.

23. Driessens, F.C.M. (1986) Thermodynamics of ionic solid solutions and its application to the formation and stability of biominerals. Ber. Bunsenges. Phys. Chem. 90, 760-763.

24. Driessens, F.C.M. (1986) Ionic solid solutions in contact with aqueous solution in Geochemical processes at mineral surfaces (eds. Davis, J.A. and Hayes, K.F.) pp 524-560. Am. Chem. Soc., Washington DC.

25. Francis, M.D. and Webb, N.C. (1971) Hydroxyapatite formation from a hydrated calcium monohydrogen phosphate precursor. Calcif. Tissue Res. 6, 335-342.

26. Brown, W.E. (1966) Crystal growth of bone mineral. Clin. Orthop. 44, 205-220.

27. Driessens, F.C.M. (1980) The mineral in bone, dentin and tooth enamel. Bull. Soc. Chim. Belg. 89, 663-689.

28. Driessens, F.C.M. and Verbeeck, R.M.H. (1982) The probable phase composition of the mineral in sound enamel and dentin. Bull. Soc. Chim. Belg. 91, 573-596.

29. Verbeeck, R.M.H., Driessens, F.C.M., Borggreven, J.M.P.M. and Wöltgens, J.H.M. (1985) Concentration gradients of some minor components in human tooth enamel. Bull. Soc. Chim. Belg. 94, 237-243.

30. Driessens, F.C.M. and Verbeeck, R.M.H. (1985) Dolomite as a possible magnesium containing phase in human tooth enamel. Calcif. Tissue Int. 37, 376-380.

31. Driessens, F.C.M., Heijligers, H.J.M., Borggreven, J.M.P.M. and Wöltgens, J.H.M. (1985) Posteruptive maturation of tooth enamel studied with the electron microprobe. Caries Res. 19, 390-395.

32. Driessens, F.C.M. (1980) Probable phase composition of the mineral in bone. Z. Naturforsch. 35c, 357-362.

33. Driessens, F.C.M. and Verbeeck, R.M.H. (1986) The dynamics of bone mineral in some vertebrates. Z. Naturforsch. 41c, 468-471.

34. Driessens, F.C.M., Borggreven, J.M.P.M. and Verbeeck, R.M.H. (1987) The dynamics of biomineral systems. Bull. Soc. Chim. Belg. 96, 173-179.

35. Driessens, F.C.M., Terpstra, R.A., Bennema, P., Wöltgens, J.H.M. and Verbeeck, R.M.H. (1987) On the possible relation between morphology and precursors of the crystallites in calcified tissues. Z. Naturforsch. 42c, 916-920.

36. Driessens, F.C.M., Verbeeck, R.M.H., Dijk, J.W.E. van and Borggreven, J.M.P.M. (1988) Degree of saturation of blood plasma in vertebrates with octocalcium phosphate. Z. Naturforsch. 43c, 74-76.

37. Driessens, F.C.M., Angmar-Mänsson, B., Heijligers, H.J.M., Whitford, G.M. and Verbeeck, R.M.H. (1987) Effects of acid-base status and fluoride on the composition of the mineral in developing enamel and dentin in the dog. J. Biol. Buccale 15, 225-228.

38. Driessens, F.C.M., Verbeeck, R.M.H., Dijk, J.W.E. van and Borggreven, J.M.P.M. (1987) Response of plasma calcium and phosphate to magnesium depletion: a review and its physio-

logical interpretation. Magnesium Bull. 9: 193-201.

39. Hay, D.I., Schluckebier, S.K. and Moreno, E.C. (1982)
Equilibrium dialysis and ultrafiltration studies of calcium and phosphate binding by human salivary proteins. Implications for salivary supersaturation with respect to
calcium phosphate salts. Calcif. Tissue Int. 34, 531-538.

40. Lagerlöf, F. (1983) Effects of flow rate and pH on calcium
phosphate saturation in human parotid saliva. Caries Res.
17, 403-411.

41. Driessens, F.C.M. and Verbeeck, R.M.H. (1980) Evidence for
intermediate metastable states during equilibration of
bone and dental tissues. Z. Naturforsch. 35c, 262-267.

42. Hoppenbrouwers, P.M.M., Driessens, F.C.M. and Borggreven,
J.M.P.M. (1987) The mineral solubility of human tooth
roots. Archs. oral Biol. 32, 319-322.

43. Borggreven, J.M.P.M., Driessens, F.C.M. and Dijk, J.W.E.
van (1986) Dissolution and precipitation reactions in human tooth enamel under weak acid conditions. Archs. oral
Biol. 31, 139-144.

44. Driessens, F.C.M., Theuns, H.M., Heijligers, H.J.M. and
Borggreven, J.M.P.M. (1986) Microradiography and electron
microprobe analysis of some natural white and brown spot
enamel lesions with and without laminations. Caries Res.
20, 398-405.

45. Driessens, F.C.M., Theuns, H.M., Heijligers, H.J.M. and
Borggreven, J.M.P.M. (1986) Microradiography and electron
probe analysis of some caries-like lesions of enamel prepared in vitro in human teeth. Archs. oral Biol. 31, 837-
840.

46. Driessens, F.C.M., Theuns, H.M., Borggreven, J.M.P.M. and
Heijligers, H.J.M. (1987) Electron microprobe analysis and
microradiography of some artificial laminated carious lesions. Caries Res. 21, 222-227.

47. Driessens, F.C.M., Theuns, H.M., Heijligers, H.J.M. and
Borggreven, J.M.P.M. (1987) Electron microprobe and microradiographic studies of carious enamel lesions covered
with dental calculus. J. Biol. Buccale 15, 183-187.

48. Patel, P.R. and Brown, W.E. (1975) Thermodynamic solubility product of human tooth enamel: powdered sample. J.
dent. Res. 54, 728-736.

49. Driessens, F.C.M., Theuns, H.M., Borggreven, J.M.P.M. and
Dijk, J.W.E. van (1986) Solubility behaviour of whole
human enamel. Caries Res. 20, 103-110.

50. Driessens, F.C.M., Dijk, J.W.E. van and Borggreven,
J.M.P.M. (1978) Biological calcium phosphates and their
role in the physiology of bone and dental tissues. Calcif. Tissue Res. 26, 127-137.

51. Verbeeck, R.M.H., Bruyne, P.A.M. de, Driessens, F.C.M.,
Terpstra, R.A. and Verbeek, F. (1986) Solubility behaviour
of Mg-containing β-Ca$_3$(PO$_4$)$_2$. Bull. Soc. Chim. Belg. 95,
455-476.

52. Hodge, H.C. and Leung, S.W. (1950) Calculus formation. J.
Periodontolog. 21, 211-221.

53. Leach, S.A. (1973) Dental calculus in Biological minerali-

zation (ed. Zipkin, I.) pp 587-606. Wiley, New York.

54. Kleinberg, I. and Jenkins, G.N. (1964) The pH of dental plaque in the different areas of the mouth before and after meals and their relationship to the pH and rate of flow of resting saliva. Archs. oral Biol. 9, 493-516.

55. Critchley, P., Saxton, C.A. and Kolendo, A.B. (1968) The histology and histochemistry of dental plaque. Caries Res. 2, 115-129.

56. Watanabe, T., Toda, K., Morishita, M. and Iwamoto, Y. (1982) Correlations between salivary protease and supragingival or subgingival calculus index. J. dent. Res. 61, 1048-1051.

57. LeGeros, R.Z. and Shannon, I.L. (1979) The crystalline components of dental calculi: human vs. dog. J. dent. Res. 58, 2371-2377.

58. Driessens, F.C.M., Borggreven, J.M.P.M., Verbeeck, R.M.H., Dijk, J.W.E. van and Feagin, F.F. (1985) On the physico-chemistry of plaque calcification and the phase composition of dental calculus. J. Periodont. Res. 20, 329-336.

59. Daculsi, G., LeGeros, R.Z., Jean, A. and Kerebel, B. (1987) Possible physico-chemical processes in human dentin caries. J. dent. Res. 66, 1356-1359.

B.R.Heywood

Calcium phosphate precipitation in liposomes

School of Chemistry, University of Bath,
Claverton Down, Bath BA2 7AY, UK

ABSTRACT

   In a number of biological systems the *de novo* formation of
calcium salts is directly associated with the biomembranes.
Indeed, some membrane constituents have been identified as
functional initiators of crystal nucleation and growth *in
vitro*. A biomimetic system has been used to investigate the
ability of specific membrane components to modulate calcium
phosphate precipitation. Intraliposomal precipitation of
calcium phosphate can be induced within phosphate-encapsulated
liposomes by ionophore-supported Ca^{2+} uptake. If these
reactions are initiated in metastable external solutions,
extraliposomal precipitation occurs when the endogenously-
formed crystals penetrate through the enclosing lipid bilayers
and seed the external solution. Certain acidic phospholipids
(phosphatidic acid; phosphatidylserine) inhibit this process
when incorporated into the liposome membranes. The ability of
these phosphatides to modulate hydroxyapatite crystal growth
relates both to their membrane orientation and the molecular
conformation of their polar head groups.

Recent Advances in the Study of
Dental Calculus

INTRODUCTION

Biological systems have evolved numerous complex strategies for regulating the nucleation and growth of biogenic crystalline phases.. However, it is evident from comparative studies of mineralized tissues that one feature is common to the continuum of regulatory mechanisms for the formation of these biominerals. The deposition of crystalline minerals is controlled in part by organic molecules interacting with the forming crystals [1]. Through these organic-inorganic interactions the allotropic form, particle size, morphology and crystallographic organization of the precipitated solids may be determined [2]. While considerable attention has focused on the ability of acidic glycoproteins to regulate crystal nucleation and growth [3,4], the efficacy of other macromolecular assemblies in controlling biomineralization has been noted [4,5].

The present study was stimulated by a general interest in the organic-inorganic interactions associated with biomineralization and more specifically by an interest in the biophysical aspects of membrane-mediated calcification. In a number of biological systems the *de novo* formation of calcium salts is directly associated with the biological membranes [6,7,8]. Indeed, specific membrane constituents have been identified as functional initiators of nucleation and crystal growth. It has been reported that stable non-dissociable calcium-phospholipid-phosphate complexes isolated from calcifying tissues can nucleate calcium hydroxyapatite crystals from metastable solutions [9]. More recently, biochemical studies have elegantly demonstrated the ionophoretic potential of calcifiable proteolipids during membrane-initiated calcification [10].

A liposome system has been developed as in *in vitro* model for examining the lipid-mineral interactions resulting from crystallization reactions within a membrane-enclosed microenvironment [11]. Phospholipids undergo spontaneous self-assembly in aqueous solutions to form stable, closed bilayer structures, *liposomes*. Since both the membrane composition and internal milieu can be tailored by experimental design, liposomes provide both a membrane-matrix and controlled reaction volume which can be exploited in fundamental and applied research of controlled crystal growth [12]. Applying the utilitarian principles of 'membrane-mimetic' chemistry [12], we have investigated calcium phosphate precipitation within anionic liposomes. Intraliposomal precipitation can be induced by encapsulating inorganic phosphate within the vesicles and then loading the inner space with Ca using cationic ionophore molecules sited in the phospholipid membrane (Figure 1) [13]. When these reactions are initiated in metastable external solutions extraliposmal precipitation occurs only when intraliposomally formed crystals penetrate through the enclosing lipid bilayers and seed the external solution phase [14]. The liposomes do not participate directly in the solid state reactions which means that molecular species with putative bioactivity can be incorporated into the lipid

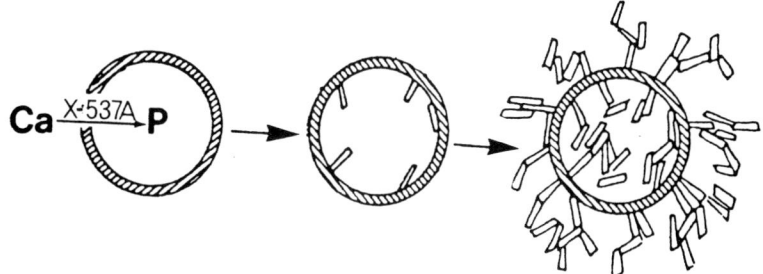

Figure 1. Schematic for ionophore-supported precipitation within phospholipid vesicles.

bilayers and their role in establishing and controlling the growth and spatial organization of crystals closely monitored. Since acidic phospholipids (APL) have been implicated in membrane-mediated calcification events, their ability to modulate apatite formation within phosphatidylcholine-rich anionic liposomes has been examined. Kinetic studies of these precipitation events indicate that while incorporated acidic phospholipids do not affect endogenous precipitation their presence in the vesicle membrane profoundly influences the subsequent exogenous precipitation reactions [15,16], Figure 2. To elucidate these findings an ultrastructural study of the calcifying liposomes was undertaken [17].

MATERIALS AND METHODS

Complementary kinetic, physico-chemical and electron microscopical techniques were used to analyse the liposome-mediated calcification events. Details of the methods employed for the preparation of the anionic liposomes, the initiation of the endogenous precipitation, the kinetic and structural analyses of the precipitation reaction and the ultrastructural examination of the lipid-mineral interactions have been published in detail elsewhere [13,14,17].

Briefly, a chloroform solution containing 7:2:1 molar mixtures of phosphatidylcholine (PC), dicetyl phosphate (DCP) and cholesterol (Chol) was evaporated to dryness and then gently resuspended in a buffered phosphate solution (50mM PIPES; 0-50mM K_2HPO_4) by mechanical agitation. Unencapsulated buffer solution was removed by gel filtration (Sephadex G50-C; 10mM HEPES) and the precipitation reactions were initiated by mixing equivalent volumes of the phosphate-encapsulated liposomes with a buffered Ca^{2+} solution (50mM HEPES; 2.25-4.50 mM $Ca(NO_3)_2$ and injecting the ionophore into this mixture (0.002-0.004 moles X-537A/mole lipid). In some experiments, the external solution phase was rendered supersaturated with the addition of inorganic phosphate (1.0-1.5mM K_2HPO_4). Liposomes incorporating other phospholipids were prepared by replacing 1-10 mole % of the 7PD:2DCP;1Chol lipid:chloroform mixture with an equivalent mole % of the APL species of

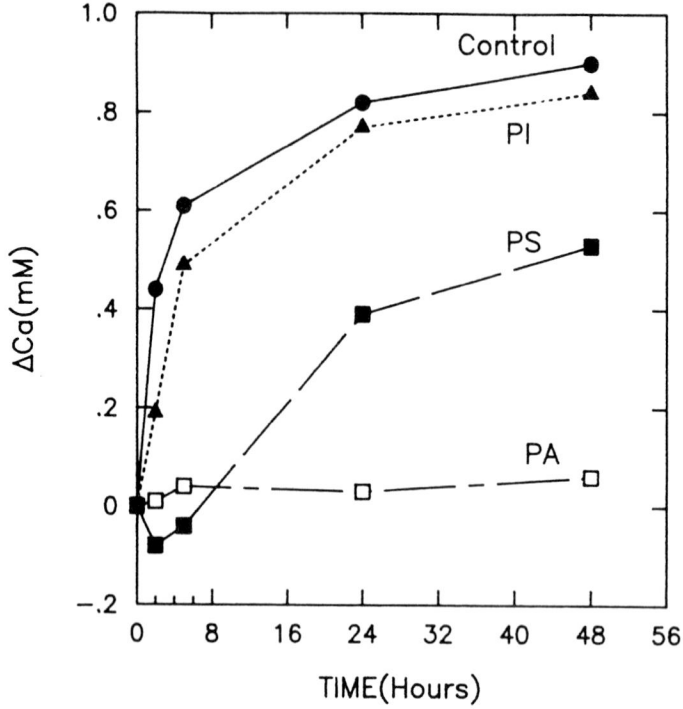

Figure 2 Effect of acidic phospholipids (10 mole %) incorporated into 7PC:2DCP:1Chol liposomes on the changes in external solution Ca^{2+} losses (Δ Ca) due to extraliposomal precipitation.

interest (phosphahatidic acid (PA), phosphatidylserine (PS), phosphatidylinositol (PI)). All experiments were carried out at 22°C, pH7.4 and 240mosm. Changes in external Ca^{2+} solution concentrations were used to follow calcium phosphate precipitation in the liposome suspensions [13]. The chemical structure of the precipitated solids was confirmed by X-ray diffraction [13].

For electron microscopy, a fixation protocol has been developed for optimal preservation of both lipid and mineral components and the mineral-membrane interactions which might arise as a result of controlled apatite formation within the liposomes [17]. Tannic acid-osmium tetroxide fixation augments the electron density of the hydrophylic layers corresponding to the polar head groups of the phospholipid molecules. Thin sections of 7PC: 1Chol liposomes so treated reveal spherical structures composed of parallel lamellae exhibiting 45Å periodicity (electron lucent regions 25-30Å thick and electron dense layers 15-20Å wide). This molecular architecture is unaffected by the precipitation reactions. The successful

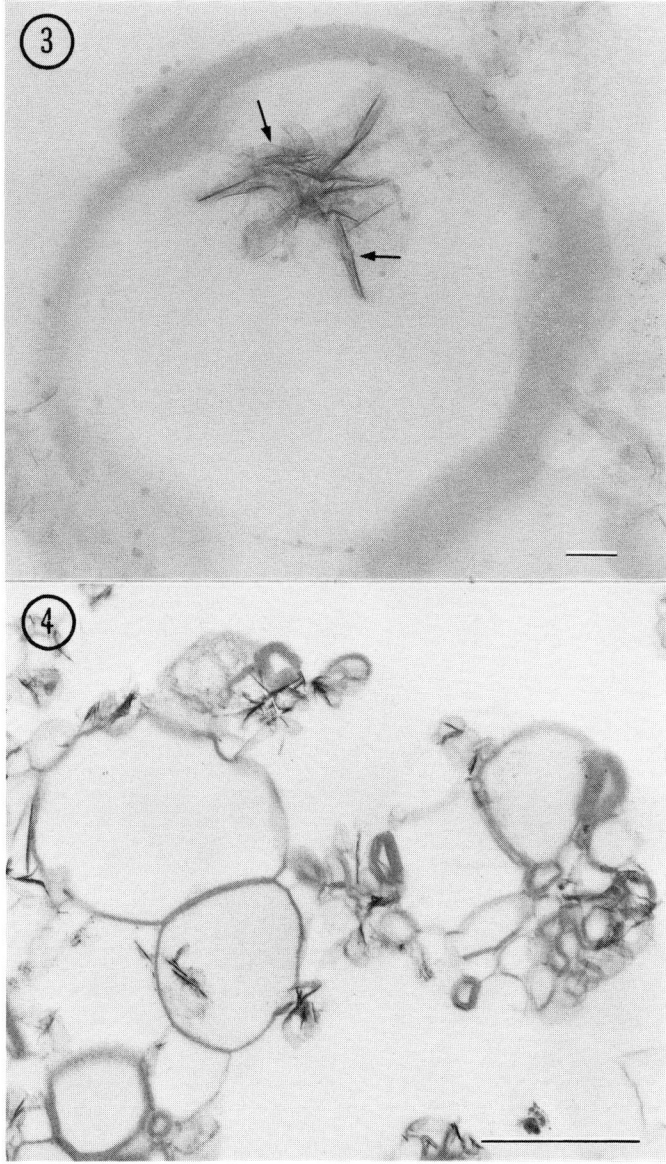

Figure 3. TEM micrograph of 7PC:2DCP:1Chol phosphate-liposomes
after X-537A mediated Ca^{2+}. Numerous electron dense
crystals (↑) of calcium hydroxyapatite are seen within the
phospholipid vesicles. Bar = 0.1 μm.

Figure 4. Extraliposomal precipitation in 7PC:2DCP:1Chol
liposomes. Crystals aggregate around the outer
surfaces of the liposomes. Bar = 1 μm.

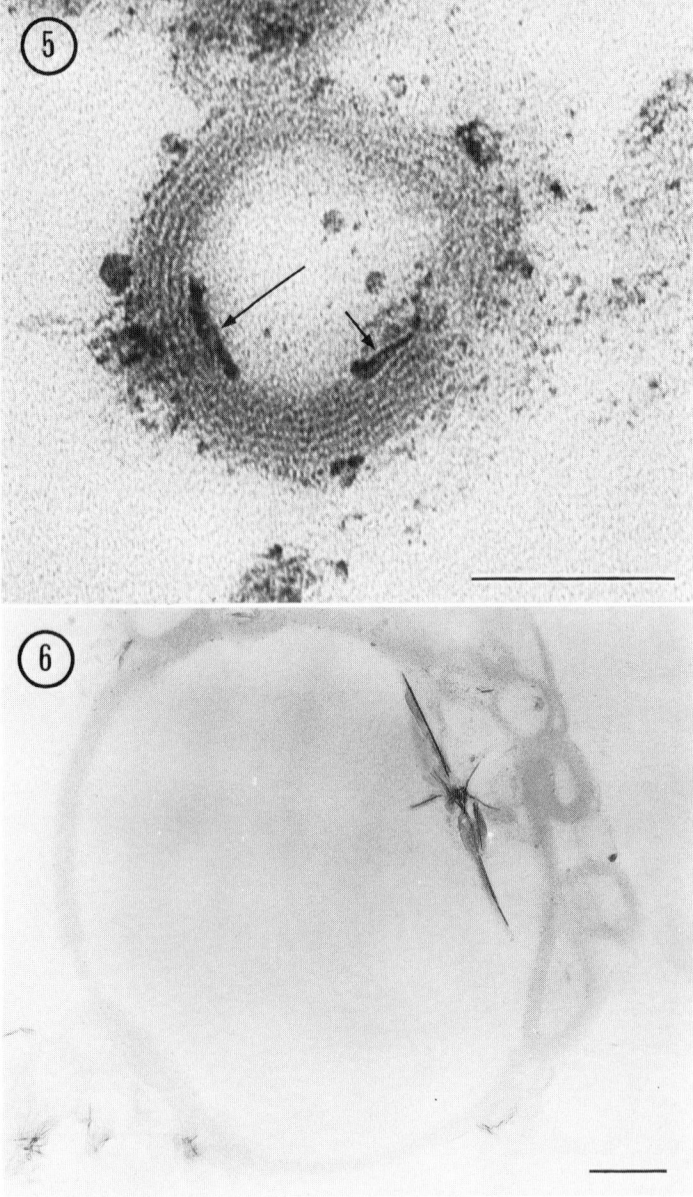

Figure 5.   Liposomes containing phosphatidic acid.   Small
            crystals adhere to the inner lipid membranes (↑).
            Bar = 0.1 μm.
Figure 6.   Intralipsomal precipitation in phosphatidylserine-
            containing liposomes.   Bar = 0.1 μm.

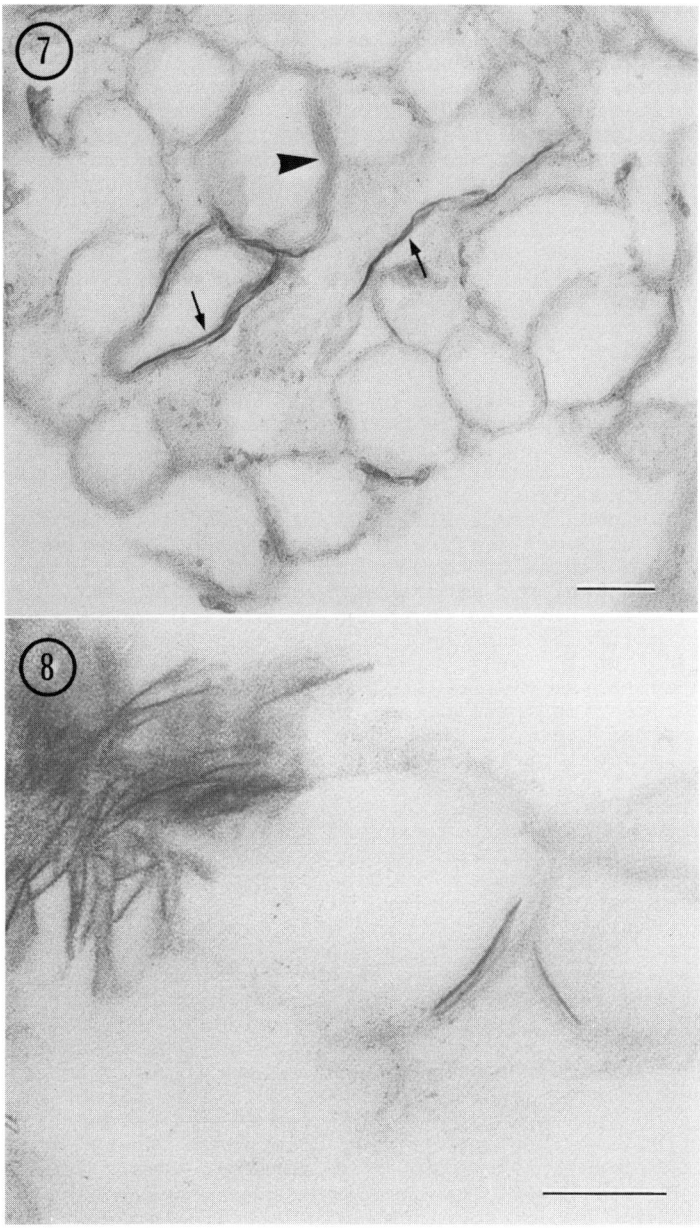

Figure 7. Extraliposomal precipitation in PS-doped liposomes. The vesicles ( ) aggregrate upon the exposed surfaces of nascent crystals (↑). Bar = 0.1 μm.

Figure 8. Calcified 7PC:2PS:1Chol (no dicetyl phosphate) liposomes. Bar = 0.1 μm.

preservation of the crystalline mineral phases is demonstrated by the presence of apatitic mineral associated within the liposomes. When the incubation conditions preclude liposomal calcification, crystalline phases are not observed.

Direct image inspection is routinely accompanied by analytical electron microscopic examination (selected area diffraction (SAD) - crystal (order) formation; X-ray microprobe analysis (EDXA) - chemical analysis). Diffraction patterns obtained from 7PC:1Chol liposomes after the induction of the precipitation events described earlier confirmed the apatitic nature of the small randomly ordered crystals while elemental analysis shows the presence of calcium and phosphorus only.

RESULTS AND DISCUSSION

Intraliposomal precipitation in phosphate-encapsulated 7PC:2DCP:1Chol liposomes is characterised by the appearance of numerous discrete mineral foci within the vesicles after calcium loading. Small (3-6nm width, 50-350nm estimated length) electron dense crystals of hydroxyapatite radiate into the aqueous lumen from their apparent point of contact with the inner lipid bilayers (Figure 3). There is no evidence for the specific orientation of these mineral deposits relative to the enclosing phospholipid membranes. When extraliposomal precipitation is induced in these anionic liposomes, aggregates of crystals form around the outer surfaces of the vesicles (Figure 4). While dynamic events cannot be visualised, a detailed appraisal of the liposomes at various time points during the reaction period clearly indicates that extraliposomal precipitation events are seeded by the exposure of intraliposomal precipitates to the external milieu. It is envisaged that extended contact between the precipitated solids and the lipid bilayers induces a transient reordering of the membrane and subsequent release of the crystals to the external solution phase. A similar mechanism is proposed by Weissman et al, [18] to explain the release of intracellular crystal deposits of pathological origin [19,20].

In contrast, intraliposomal precipitation in liposomes with 10% phosphatidic acid incorporated in the vesicle membrane is characterised by the presence of small crystals lying in close contact with, and parallel to the inner lipid bilayers (Figure 5). Only a few isolated crystals adhering to the outer surfaces of the clustered liposomes can be identified after incubation in metastable solutions. Kinetic analyses of the precipitation reactions confirm the inhibitory effect of the monoester phosphate lipid upon calcium phosphate deposition, whilst parallel experiments using diester phosphate lipids (10% phosphatidylinositol) suggest that a strong adsorption between the nascent apatite and the monoester phosphate moieties in the lipid bilayers serves to block potential growth sites on the crystal surface and thereby restricting them from seeding further precipitation [16].

The ability of the membrane constituents to modulate crystal growth relates not only to the molecular conformation of their polar head groups but also to the orientation of incorporated species within the liposome membrane.

When phosphatidylserine (PS) (10%) is incorporated into the outer lipid bilayer, intraliposomal precipitation is directly comparable with the control system (Figure 6), whereas PS-induced aggregation of the liposomes around the nascent mineral reduces the amount of precipitate seeded by intraliposomally-formed apatite (Figure 7). If this amino phospholipid is localised in the inner membrane, extraliposomal precipitation is restored to normal yields but an extended 'lag' period occurs before the onset of external precipitation [15]. Close examination of the lipid-mineral associations in the latter system (7PC:1PS:1Chol, no DCP) reveals small intraliposomal crystallites closely aligned with the membrane-aqueous interface (Figure 8).

Little is known about nucleation *in vivo* on lipid bilayer assemblies. *In vitro* studies have shown that crystal nucleation and growth can be influenced by synthetic vesicles [15,16,21]. In the present study the use of phospholipid vesicles for investigations into biomembrane-mediated calcific-ation underlined the importance of substrate identity in regulating mineral deposition within sequestered microenvironment. The nucleation and growth of calcium hydroxyapatite was markedly influenced by the incorporation of acidic phospholipids into the limiting bilayer of the liposomes. Specifically, they retarded exogenous crystallization by affecting the interaction of the bilayer with the nascent crystals. It seems likely that electrostatic interactions between membrane-associated phosphate groups and Ca^{2+} in the hydroxyapatite crystal are responsible for these effects. Controlled crystallization experiments with two-dimensional Langmuir monolayers have indicated that the co-operative influences of electrostatic and stereochemical factors is a key factor in determining the surface-environment interactions which control crystal growth [22,23]. It is hoped that the application of high resolution transmission electron microscopy techniques to the calcifying liposomes will highlight the importance of these factors in the present system.

The above studies indicate the potential of biomimetic systems for probing the molecular interactions at the organic-inorganic interface which control the biological process of controlled crystallization. It is envisaged that in future research the system will be used for further studies directed towards an understanding of the influences of macromolecular assemblies upon crystal nucleation and growth in order to better define the degree to which interactions between the organic and inorganic components affect biological crystal growth. Significant advances have already been made in such areas as reactivity control, catalysis, drug targeting and semi-conductors through the use of such membrane-mimetic systems. It is to be hoped that information gained from liposome-mediated crystallization experiments might be exploited in the development of novel strategies for the control, management and repair of pathological disorders associated with abnormal mineralization.

ACKNOWLEDGEMENTS

The author is indebted to Dr E.D. Eanes for many valuable discussions and suggestions during the course of this work.

B.R.H. was a post-doctoral fellow at the National Institutes of Health, Bethesda, MD 20892, USA for the period covered by these investigations.

REFERENCES

1. Weiner, S. (1986) CRC. Critical Reviews in Biochemistry, 20, 365-408.
2. Mann, S. (1988) Nature, 332, 119-224.
3. Addadi, L. and Weiner, S. (1986) Mol. Cryst. Liq. Cryst., 134, 305-332.
4. Fisher, L.W. and Termine, J. (1985) Clin. Orthopaedics, 200, 363-385.
5. Boskey, A.L. (1978) Metab, Bone Dis. Rel. Res., 1, 137-142.
6. Vogel, J.J., Boyan-Slayers, B.D. and Campbell, M.M. (1978) Metab. Bone Dis. Rel. Res., 1, 149-153.
7. Westbreok. P., de Jong, E.W., van der Wal, P., Borman, A.H., de Vrind, J.P.M., Kok, D., de Bruijn, W.C. and Parker, S.B. (1984) Phil. Trans. Roy. Soc., B 304, 435-444.
8. Markel, K., Roser, U. Mackenstadt, U. and Klosterman, M. (1986) Zoomorphology, 186, 232-243.
9. Boskey A.L. and Posner, A. (1977) Calcif. Tiss. Res., 23, 251-258.
10. Swain, L.D. and Boyan, B.D. (1988) J. Dent. Res., 67, 526-530.
11. Eanes, E.D. and Heywood, B.R. (1988) in 'Liposome Technology in Biomineralization Research' (ed. H. Myers) Karger Basel, Switzerland, in press.
12. Fendler, J.H. (1980) Acc. Chem. Soc., 13, 7-13.
13. Eanes, E.D., Hailer, A.W. and Costa, J.L. (1984) Calcif. Tiss. Int., 36, 421-430.
14. Eanes, E.D. and Hailer, A.W. (1985) Calcif. Tiss. Int., 37, 390-394.
15. Eanes, E.D. and Hailer, A.W. *1987) Calcif. Tiss. Int., 40, 43-48.
16. Eanes, E.D., Hailer, A.W. and Heywood, B.R. (1988) Calcif. Tiss. Int., 43, 226-230.
17. Heywood, B.R. and Eanes, E.D. (1987) Calcif. Tiss. Int., 41, 192-201.
18. Weissmann, G. and Rita, G.A. (1972) Nature, 240, 167-172.
19. Elferink, J.G.R. (1986) Biochem. Med. Met. Biol., 36, 25-35.
20. Schumacher, H.R., Fishbein, P., Phelps, T., Tse, R., and Kruser, W. (1978) Arthritis Rheum, 18, 783-792.
21. Mann, S. and Hannington, P. (1988) J. Colloid Interface Sci., 122, 326-340.
22. Landau, M., Levanon, M., Lieseromitz, L., Lahav, M. and Sagiv, J. (1985) Nature 318, 353-356.
23. Mann, S., Heywood, B.R., Rajam, S. and Birchall, J. (1988) Nature, 334, 62-65.

B.D.Boyan[1],
L.D.Swain[1]
A.L.Boskey[2]

Mechanisms of microbial calcification

[1]The University of Texas Health Science Center at San Antonio, San Antonio, Texas 78284, USA and [2]Hospital for Special Surgery, New York City, New York 10021, USA

ABSTRACT

Calcification of dental plaque has been associated with proteolipids (Pr) which are integral membrane proteins that are soluble in organic solvents. In microorganisms which are able to support initiation by hydroxyapatite formation, the process appears to be proteolipid-dependent. Ca and inorganic phosphate (P_i) interact with phosphatidyserine structured by the proteolipids in the membrane to form calcium:phospholipid:phosphate complexes (CPLX) which then support subsequent hydroxyapatite deposition. Experiments using deposition Bacterionema matruchotii as model suggest that some, but not all, of the proteolipids in the bacteria are involved in the process and these can be co-isolated with CPLX (Pr-CPLX). Ability of oral bacteria to calcify is based on the biochemical composition of Pr, its concentration in the membrane, and its associated phospholipids. These characteristics may change as the culture matures. Events at the initiation site include the import of Ca and P_i and the export of protons in a non-energy dependent manner through specific proteolipid ion channels. The process can be modulated metabolically and by altering the mineral composition of the medium.

Recent Advances in the Study of
Dental Calculus

Role of bacteria in calculus formation

Calcification of dental plaque appears to be associated with oral bacteria. It is time dependent (1), suggesting that there is a maturation of the plaque that must occur before calcification can takes place. In addition, mineral deposition appears to occur initially in discrete foci (2), suggesting that some, but not all, bacteria can serve as initiation sites. A survey of the ability of isolated microorganisms to support in vitro calcification (36) supports this hypothesis because not all bacteria become calcified when incubated in vitro under conditions which simulate the oral cavity.

Recent studies have shown that hydroxyapatite formation in Bacterionema matruchotii depends on the presence of specific membrane proteins called proteolipids (7). These are a class of integral membrane proteins that are associated with phospholipids and are involved in ion transport (8-10). While all microorganisms have proteins of this type, not all microorganisms calcify, suggesting that there may be distinct differences in the biochemical characteristics of these proteins, in their concentration, in their metabolism, or in other related membrane components.

Calcification of dental calculus also is dependent on proteolipids (11). When demineralized calculus is extracted with chloroform:methanol 2:1, the extract supports hydroxyapatite formation in vitro, whereas the residue does not. Of the material extracted by the organic solvent, only proteolipids and their associated phospholipid retain the ability to support hydroxyapatite formation in metastable calcium phosphate solution (MCPS). It is probable that this proteolipid is derived from the bacterial plaque.

Diversity in bacterial proteolipids isolated from oral bacteria has been verified in a number of studies. Howell and Boyan (12) demonstrated that proteolipids isolated from Actinomyces naeslundii, which do not calcify in culture, differ from those isolated from B. matruchotii, both in concentration and in composition. Similarly, proteolipids isolated from Streptococcus sanguis I, which does not calcify, have a different composition from proteolipids isolated from S. sanguis II and S. mitis which do calcify in culture (13).

The concentration of proteolipids also appears to be a factor. The calcifiability of a microorganism correlates positively with increasing proteolipid content (13,14). This is the case for cultures of B. matruchotii as well (15). As cultures age, the content of total proteolipid increases, peaking in cultures which are 8 days old. This is coincident with the appearance of hydroxyapatite in the bacteria (16). The increase in proteolipid content, specifically of proteolipid with an apparent molecular weight of 10,000 (15), may be due to increased synthesis, to changes in the membrane environment of the protein thereby altering the dynamics of its isolation, or to specific degradation of other membrane components. Whatever the cause, it may explain the observation that some maturation of plaque occurs before calcification is seen.

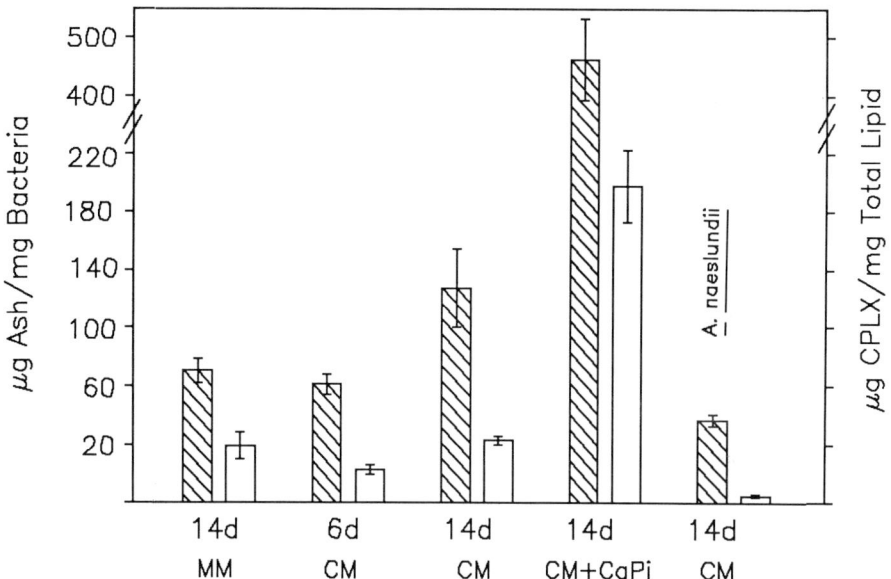

Figure 1. Comparison of ash content (hatched bars) and CPLX content in <u>Bacterionema</u> <u>matruchotii</u> incubated for 14d in maintenance medium (MM), calcification medium (CM), or CM replenished with ca and phosphate (CaP$_i$) daily. <u>Actinomyces naeslundii</u> was analyzed for comparison. Data are adapted from Boyan-Salyers and Boskey, 1980. Values represents mean ± S.D.

Role of lipids in microbial calcification

Studies in our laboratory using immunoelectron microscopy have shown that proteolipids in <u>B</u>. <u>matruchotii</u> are present in both the outer membrane and inner mesosomal membranes (3). Mineralization in these bacteria is proteolipid-dependent; only fractions of the bacteria which contain proteolipids support <u>in</u> <u>vitro</u> hydroxyapatite formation (7). While both the protein and its associated lipid are required for mineral formation in MCPS (17), it is becoming clear that these phospholipids actually provide the sites for initial mineral formation. Boskey and her colleagues (18) had shown the presence of calcium:phospholipid:phosphate complexes (CPLX) in tissues which were calcified, or about to calcify, and noted their absence from tissues were not calcified. Together with Boyan (19) she observed that CPLX were present in <u>B</u>. <u>matruchotii</u> whether or not they were cultured in calcification permissive medium. CPLX concentration increased when the bacteria were cultured in calcification medium, particularly when the medium was replenished with Ca and inorganic phosphate (P$_i$) daily. However, CPLX were absent from bacteria like <u>A</u>. <u>naeslundii</u> which did not calcify in culture under these conditions (Figure 1).

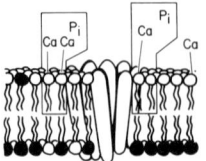 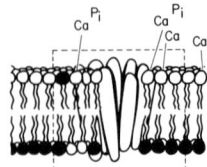 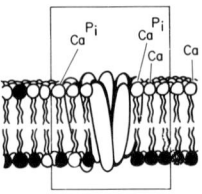

Ca-Phospholipid-Phosphate
(Boskey and Posner, 1976)
No Demineralization
Sonication
Ethanol:Ether Precipitation

Proteolipid
(Ennever et al, 1976)
2N Formic Acid
No Sonication
NaCl Wash
Acetone Precipitation
Gel Filtration

Ca-Proteolipid-Phosphate
(Boyan and Boskey, 1982)
50 mM Phthalic Acid, pH 5.5
No Sonication
NaCl Wash
Acetone Precipitation
Gel Filtration or Ether Precipitation

Figure 2. Comparison of isolation protocols for CPLX, proteolipid, and proteolipid-CPLX. In all cases the solvent used for extraction was chloroform:methanol 2:1. Differences are described in detail iln the appropriate references.

These observations suggested that there might be a specific relationship between proteolipids and CPLX. As shown in Figure 2, the extraction protocol used in Boyan's (7) and Boskey's (20) laboratories yielded isolated proteolipids and CPLX respectively. By developing methods for co-isolation (21), these investigators were able to show that CPLX associated with proteolipid was similar to that isolated independently from the bacteria and to CPLX isolated from vertebrate tissues which form hydroxyapatite.

These data suggested that CPLX formation occurred on the membrane surface and involved phosphatidylserine specifically associated with the proteolipid. This hypothesis was further supported by the fact that phosphatidylserine is the predominant phospholipid in CPLX and in the 10,000 M_r proteolipid isolated from B. matruchotii (22). This phospholipid is relatively rare in B. matruchotii since it is rapidly metabolized to phosphatidylethanolamine making its enrichment in CPLX and the proteolipid extract more notable. One reason that phosphatidylserine might be selected for by the 10,000 M_r proteolipid and CPLX may be that it is inaccessible to enzymes either due to the presence of protein or to binding of Ca.

It is now clear that those proteolipids isolated from B. matruchotii using sequential extraction of the bacteria with chloroform:methanol 2:1 and chloroform:methanol:HCl 200:100:1 are not strictly homologous which those proteins co-isolated with CPLX. Ennever et al. (7) have shown that proteolipid isolated by chloroform:methanol 2:1 alone is sufficient to support hydroxyapatite formation. Proteolipids isolated in this manner exhibit a single band with a relative molecular weight of 10,000 on SDS-ployacrylamide gel electrophoresis (unpublished data) and form CPLX in vitro (19) prior to forming hydroxyapatite, suggesting that a 10,000 M_r protein would be present in proteolipids co-isolated with CPLX. In order to support ion transport, however, the 10,000 M_r proteolipid must be recombined with those proteins isolated with chloroform:methanol:HCl 200:100:1, indicating that the complete function of proteolipids in biologic calcification relies on its structural characteristics in the membrane and the formation of multiprotein ion channels.

As shown in Figure 3, proteolipids co-isolated with CPLX (Lane 3) are not strictly homologous with those isolated independently by sequential chloroform: methanol and acidified chloroform: methanol extraction (Lane 2). Most striking is the absence of a 10,000 M_r protein in Pr-CPLX co-isolates.

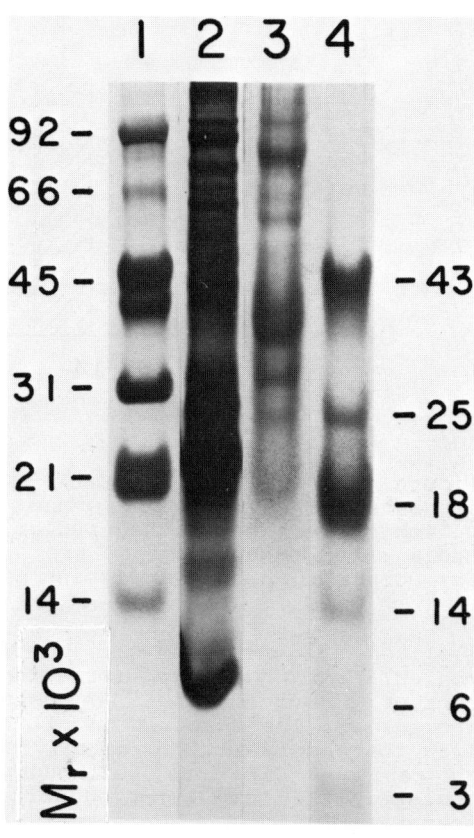

Figure 3. SDS-poly-acrylamide gel electro-phoretogram showing molecular weight standards (Lane 1 and 4) proteolipid (Lane 2) and PR-CPLX (Lane 3). Stained with silver.

Proteolipids tend to behave anomalously on gels and to aggregate as higher molecular weight oligomers and multimers. Thus, it is possible that formation of CPLX selects for, or "captures", specific multimeric functional units. In fact, this appears to be the case. Antibodies generated to the 10,000 M_r proteolipid recognizes several higher M_r protein bands in Western blots of Pr-CPLX (unpublished data).

Metabolic regulation
The ability of oral bacteria to calcify can be regulated by changes in phospholipid metabolism (22) and by the ion composition of the medium in which they are incubated (23). As shown in Figure 4, when B. matruchotii are incubated in calcification permissive medium the amount of CPLX and Pr-CPLX

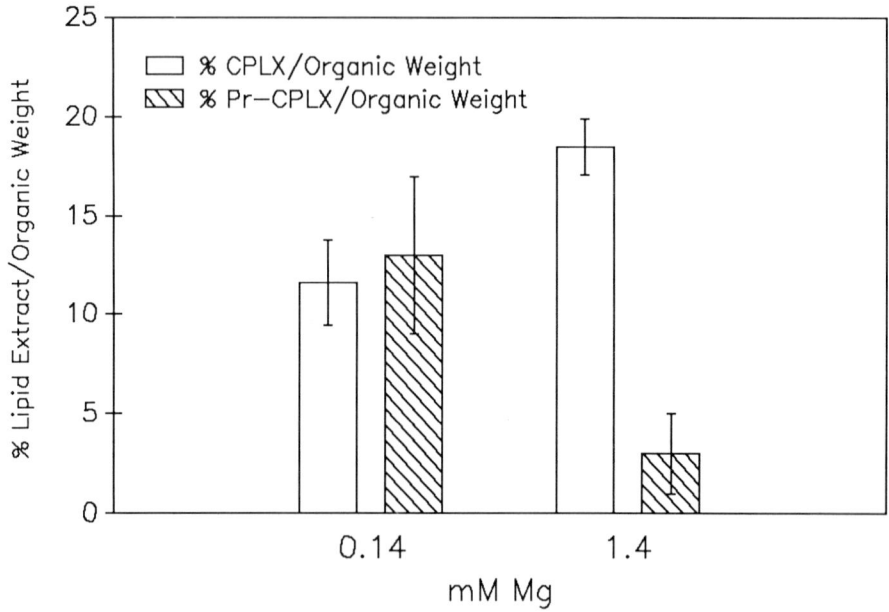

Figure 4.   Content of CPLX amd Pr-CPLX in B. matruchotii
incubated in calcification permissive medium (1mM Ca, 3.9mM P_i,
0.14mM Mg) or Mg enriched medium (1mM Ca, 3.9mM P_i, 1.4mM Mg).
Media salt concentrations were adjusted so that ionic strength
was comparable. Data represent means ± S.D.

isolated are comparable. However, when Mg is added to the
medium, hydroxyapatite formation is inhibited, and Pr-CPLX
content is decreased. In contrast the amount of CPLX is
increased. The composition of this CPLX is markedly different
from that isolated from calcifying bacteria (unpublished data).
Mg replaces Ca; thus, although mineral-complexed lipids are
isolated, they are non-productive in supporting hydroxyapatite
formation.

Summary

    These data suggest a model of the hydroxyapatite initiation
site. Proteolipids participate in ion transport by forming
specific ion channels. These may promote import of Ca and P_i
and export of H^+ released during the formation of the
hydroxyapatite crystals. At least one component of the ion
channels, the 10,000 M_r proteolipid, is specifically associated
with phosphatidylserine which then interacts with Ca and P_i to
form CPLX. Once CPLX has formed, apatite deposition follows
when sufficient Ca and P_i are present and the concentration of
inhibitors is low. Thus, B. matruchotii appears to be primed
for hydroxyapatite formation whereas A. naeslundii is not
(13,19).

Acknowledgement
These studies were supported by PHS Grant DE 05932. The authors wish to acknowledge the assistance of Linda Keller, Robert Dennis, Linda Kain and Ruben Gomez.

REFERENCES

1.   Mandel, I. et al. (1957) J. Periodont., 28, 132-137.
2.   Listgarten, M. (1976) J. Periodont., 47, 1-18.
3.   Boyan, B., Landis, W., Knight, J., Dereszewski, G. and Zeagler, J., (1984) Scanning Electron Microscopy, 4, 1793-1800.
4.   Streckfuss, J., Smith, W., Brown, L. and Campbell, M., (1974) J. Bacteriol., 120, 502-506.
5.   Streckfuss, J., Smith, W., Brown, L. and Campbell, M. (1979) J. Dent. Res., 58, 1916-1917.
6.   Ennever, J., Vogel, J. and Brown, L. (1972) J. Dent. Res., 51, 1483-1486.
7.   Ennever, J., Vogel, J. and Brown, L. (1972) J. Dent. Res., 51, 1483-1486.
8.   Folch-Pi, J. and Stoffyn, P. (1972) Ann. N.Y. Acad. Sci., 195, 86-107.
9.   Schlesinger, M. (1982) Ann. Rev. Biochem., 50, 193-206.
10. Boyan, B. and Clement-Cormer, Y. (1984) In: Membranes, Detergents, and Receptor Solubilization, Alan R. Liss, Inc., pp. 47-63.
11. Ennever, J., Vogel, J., Boyan-Salyers, B. and Riggan, L. (1979) J. Dent. Res., 58, 619-623.
12. Howell, R. and Boyan-Salyers, B. (1980) J. Dent. Res., 59, 1999-2005.
13. Boyan, B., Swain, L. and Gomez, R. (1986) In: The Sixth International Symposium on Biomineralization. (R. Crick, ed.).
14. Khare, A., Swain, L., Gomez, R. and Boyan, B. (1988) J. Dent. Res., IADR Abstract.
15. Boyan, B., Dereszewski, G., Hinman, B., Florence, M. and Griffiths, G. (1982) In: Fifth International Workshop on Calcified Tissues, Kiryat Anavim, Israel, (Sela, J. ed.), Excerpta Medica, pp. 12-17.
16. Boyan-Salyers, B., Vogel, J. and Ennevr, J. (1978) J. Dent. Res., 57, 291-295.
17. Ennever, J. and Vogel, J. (1980) J. Dent. Res., 59:1175.
18. Boskey, A., Goldberg, M. and Posner, A. (1977) Proc. Soc. Exp. Biol. Med., 157, 588-591.
19. Boyan-Salyers, B. and Boskey, A. (1980) Calcif. Tissue Int., 30, 167-174.
20. Boskey, A. and Posner, A. (1976) Calcif. Tissue Res., 19, 273-283.
21. Boyan, B. and Boskey, A. (1984) Calcif. Tissue Int., 36, 214-218.
22. Boyan, B., Anderson, G., Dereszewski, G., Howell, R. and Rapley, J. (1983) In: Calcium Binding Proteins. (B. deBernard et al., eds.) Elsevier Science Publishers, pp. 11-18.

L.D.Swain
B.D.Boyan

Ion-transport properties of membrane proteins associated with microbial calcification

The University of Texas Health Science Center at San Antonio, San Antonio, Texas 78284, USA

ABSTRACT

The study of calcium hydroxyapatite deposition has led to the hypothesis that specific proteolipids facilitate calcium hydroxyapatite formation via their role in ion translocation. The studies utilized the proteolipid-dependent calcification of Bacterionema matruchotii as a model of membrane facilitated mineralization. Preliminary results indicated that calcifiable proteolipid enhances proton flux through the membranes of proteoliposomes. Proteolipids serve two functions in the bacterial membrane. One function is to facilitate formation of CPLX. The other is to form specific ion channels. Particularly important is the import of Ca and the export of protons. At least one component of B. matruchotii proteolipid binds DCCD, a characteristic of ion translocating proteolipids. Our studies indicate that proteolipids which do not bind DCCD are also necessary to form ion channels. At the time crystals are formed, ATP has been depleted. Therefore non-energy dependent ionophores, and not ion pumps per se, are prerequisites. Proteolipid enhancement of H^+ transport when reconstituted into proteoliposomes is lost when either neutral extracted proteolipid or acidified extracted proteolipid is used alone, suggesting that at least two proteins are necessary. Enhancement of H^+ transport is abolished when ^{14}C-DCCD-proteolipid conjugates are used. Ion transport is necessary in vivo where export of H^+ from a nucleating site is as critical as transport of Ca and P_i to the site. However, in vitro, where any H^+ produced is buffered or immediately diluted, the presence of a complex nucleation site may not be necessary. This appears to be the case because the 10K component of proteolipid is sufficient to support hydroxyapatite formation in metastable calcium phosphate solution. Enrichment of proteolipid in the organic matrices of ectopic calcified deposits underscores the special relationship of these membrane components in mineralization.

INTRODUCTION

Initial deposition of hydroxyapatite (HA) has been associated with membranes or their constituents in calcifying bacteria (1,2), dentine (3), bone (4), and during endochondral ossification (5). The classic morphologic demonstration of this close association of membrane and mineral is in electron micrographs of growth cartilage (6). Similar observations have been made in mantle dentine (3) and fracture callus (7), where deposition of mineral is regulated, and in ectopic calcifications such as dental calculus (8,9).

Studies using bacterial calcification as a model of membrane-mediated HA formation indicate that calcification is the result of specific membrane characteristics, particularly the structuring of phosphatidylserine and protein such that subsequent interaction with Ca and Pi results in CPLX formation (10). Calcifiable Pr and CPLX have been co-isolated from calcifying bacteria (11).

Calcification of _Bacterionema matruchotii_ provides a particularly useful tool since hydroxyapatite formation is proteolipid-dependent, involving formation of CPLX, and subject to cellular regulation and the sequence of mineral deposition has been defined for both calcifying and non-calcifying cultures. _B. matruchotii_ is part of the calculus forming biota and is a component of the "corn cob" structure in maturing dental plaque.

ASPECTS OF MEMBRANES

In a survey of 14 oral bacteria, Ennever (12) found that only 4 calcified in culture. This indicates that only a few, but not all, biological membranes are suitable for initiating HA crystal formation. For example, although isolated mitochondria can accumulate large calcium and phosphate pools and form insoluble aggregates, crystalline HA is very rarely observed (13). When various lipid extracts from _B. matruchotii_ are incubated in matastable calcium phosphate solutions, CPLX formation occurs predominantly on proteolipid-associated phospholipids (10,11). Phospholipids not associated with calcifiable proteolipid neither form CPLX, nor do they support HA deposition (11). Thus, membranes that calcify are in some way different from the majority of biological membranes.

The importance of lipids in the process of dental calculus formation was first recognized by Mandel and Levy (14). Phosphatidylserine and the phosphatidylinositides, especially phosphatidylinositol-4-phosphate and phosphatidylinositol-4,5-bisphosphate, are enriched in CPLX and calcifiable proteolipid extracts from a variety of calcifying tissues (11) suggesting that these phospholipids have an important role in function. However, the flow of ions across cell membranes is orders of magnitude faster than is predicted by the solubility constants of ions in lipids (15). This suggest that ions are translocated across biological membranes through discontinuities in the lipid bilayer (16). These discontinuities are of course proteins. It has been shown in fact that a major function of these integral membrane proteins is the generation of aqueous transmembrane channels that permit the rapid diffusion of small hydrophilic molecules (17).

Proteolipid Composition.

Proteolipid proteins are an ubiquitous class of integral hydrophobic membrane proteins (18). Even though in different membranes these proteins appear to be unique, they in fact share important physical characteristics such as solubility in organic solvents, insolubility in aqueous solvents and resistance to proteolytic enzymes (19). They are hydrophobic, can be converted to solubility in aqueous solvents only with considerable difficulty (20), and tend to be strongly associated with acidic phospholipids, particularly PS and the phosphatidylinositides (PI) (20-22). Addition of Pr to mixed lipid liposomes reorders these lipids into discrete functional domains about the protein core (23). The amphipathic nature of these proteolipids is demonstrated in their conformational flexibility, which may relate to their capacity to form ion channels in phospholipid bilayers (24).

Previous studies in our laboratory have shown that in vitro calcification is proteolipid-dependent yet the mechanism by which proteolipid facilitates membrane-mediated hydroxyapatite formation is not known. In the prokaryote model, calcium and phosphate interact with phosphatidylserine that has been structured in the membrane by proteolipid to form calcium:phospholipid:phosphate complexes (CPLX) which can then serve as a template for hydroxyapatite. At lease one or more proteolipids present in the bacterial membrane are conserved in CPLX. It is known that hydroxyapatite formation requires transport of ions to the site of nucleation and export of protons from those sites in order to maintain optimal pH for crystal growth. Since many proteolipids investigated in other tissues and cell types have a functional role in ion transport, we hypothesized that calcifiable proteolipid might have a similar function.

Proteolipid Characteristics.

Formation of HA in biological systems involves ion flux to and from the nucleating site. Our results suggest that calcifiable Prs are involved in ion transport (25). BMPr binds the H^+ channel inhibitor dicyclohexylcarbodiimide (^{14}C-DCCD). Competitive binding with DCCD, ethyldimethylaminopropyl-carbodiimide (EDAC), or glycine ethyl ester demonstrates that the binding is specific and to a carboxyl group in the hydrophobic region of the membrane. Label was associated with a singe Pr of 8,500 M_r. Both rate and extent of H^+ translocation in bacteriorhodopsin (BRh) liposomes were enhanced by increasing amounts of BMPr. DCCD abolished BMPr augmented ion transport in a concentration dependent manner.

We investigated the use of proteoliposomes to study the role of calcifiable proteolipids in ion transport. The fact that these proteolipids bind DCCD suggests that they are involved in ion flux. Accordingly we developed technology using bacteriorhodopsin proteoliposomes. Bacteriorhodopsin (BRh)

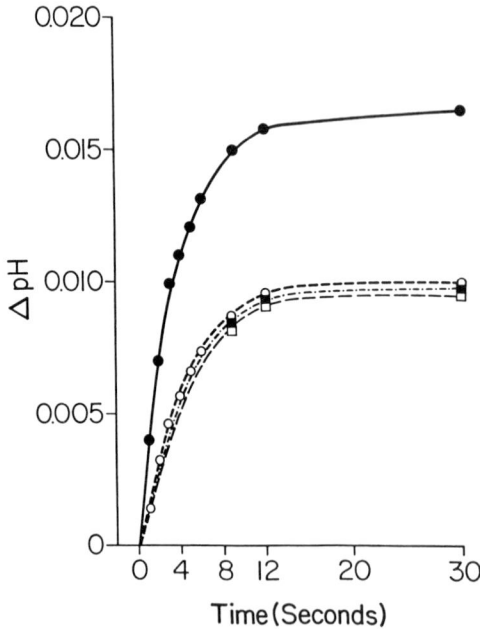

Figure 1. The Effect of calcifiable proteolipid on the kinetics of proton translocation in liposomes containing bacteriorhodopsin. Liposomes containing combined proteolipid extracts (solid circles) are compared to liposomes containing only BR (open circles); liposomes containing BR plus neutral extract proteolipid solid squares); and liposomes containing BR plus acidified extract proteolipid.

generates an electrochemical proton gradient across the cell membrane by operating as a light driven proton pump (26). Proteoliposomes containing BRh were constructed from soybean asolectin by a cholate dialysis technique modified by Lind (27). Baseline parameters were determined on BRh-liposomes that contained no calcifiable proteolipid. Both the rate and extend of proton pumping were determined per BRh molecule, and the ability of calcifiable proteolipid to augment transport was assessed as the enhancement of BRh-induced proton pumping. To ensure that any enhancement of BRh-induced translocation was due to proteolipid and not to an interaction between BRh and proteolipid, proteoliposomes were also constructed that contained proteolipid-DCCD conjugates. Proteolipid increases the rate and extent of proton translocation, whereas proteolipid-DCCD conjugates do not, indicating that DCCD blocks proteolipid enhancement of bacteriorhodopsin proton pumping. Proteolipid acts in a way indicative of ionophoric activity.

Proteolipid function.

BMPr enhancement of H^+ transport when reconstituted into BRh proteoliposomes is lost when either the neutral extracted Pr or acidified extracted. Pr is used alone (28) suggesting that at least two Prs are necessary (Figure 1). The Prs appear to act as true ionophores, such as valinomycin (29). Enhancement of H^+ transport is abolished when ^{14}C-DCCD-Pr conjugates are used. Ion transport is necessary in vivo where export of H^+ from a nucleating site is as critical as transport of Ca and P_i ions to the site. However in vitro, where any H+ produced is buffered or immediately diluted, the presence of a complex nucleation site may not be necessary. This appears to be the case because the 10K component of BMPr is sufficient to support HA formation in metastable calcium phosphate solution.

That ion transport by B. matruchotii proteolipid requires at least two or more proteolipid subunits is consistent with the literature. In fact the conformation, hydrophobicity profiles, amino acid compositions of these proteins have many similarities and all of the channel proteins are composed of multiple subunits that aggregate in order to conduct ions (30). For example, proton translocation occurs at the interphase between different F_o subunits (31).

Experiments described above demonstrated that calcifiable Prs can function as ionophores. To eliminate confusion in interpreting the data due to the presence of bacteriorhodopsin in the liposomes, we have modified the black lipid membrane for use in our system. We have built an apparatus which enables us to precisely determine both the phospholipid and Pr composition of the lipid bilayer across a microaperature in a teflon membrane. Solution characteristics are varied on either side of the membrane, and the movement of ions across the bilayer is monitored. We have completed preliminary experiments using soybean asolectin phospholipids which verify that it will be suitable for our proposed studies. With the asolectin membranes formed in these trials, the average life of the membranes formed was 72 minutes, a time period sufficient for ion transport studies to be conducted.

Initiation Site.

A mechanism for export of H^+ is as important as mechanisms for import of calcium and phosphate. The model shown in Figure 2 illustrates the information summarized in this paper and allows for the formulation of additional concepts. The first hydroxyapatite crystals formed are on the inner leaflet of the bacterial membrane and thus one would predict the M_r 10,000 proteolipid to be localized here. Both the M_r 10,000 and 8,500 DCCD-binding proteins are required for translocation activity therefore suggesting that they span the membrane. Our results do not indicate what ions are transported by calcifiable proteolipid however, as indicated by the model, a calcium-proton exchanger is a strong possibility. Researchers (32) have described a calcium-proton antiporter in Escherichia coli which exchanges H^+ for calcium and Tsujibe and Rosen (33) and Ambudkar et al.(34) have shown that antiporter system to be sensitive to DCCD.

It has been demonstrated that Na^+-H^+ exchangers mediate the uphill extrusion of H^+ coupled to, and thus energized by, the downhill entry of Na^+ (35). The uptake of Na^+ is electroneutral either in a presence or absence of a H^+ gradient, indicating that a fixed 1:1 stoichiometry exists for the exchange process.

As a consequence of its mediating H^+ transport, the Na^+-H^+ exchanger therefore plays an important role in the regulation of intracellular pH (36). Since in vivo mineralization requires a sensitive pH control mechanism, and because the Na^+-H^+ exchanger system is a common mechanism of pH control, one is well justified to investigate the possibility that such a system might be involved in the control of H^+ flux associated with HA

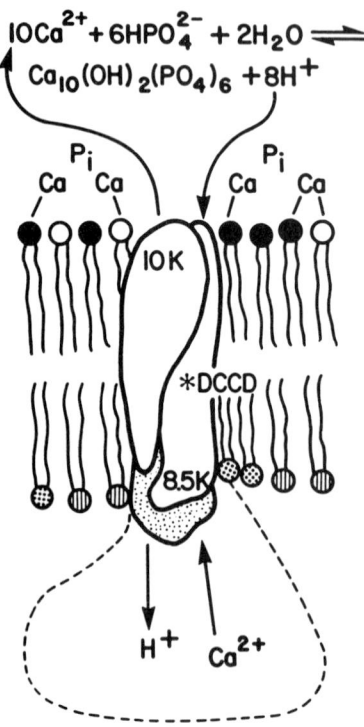

$$10Ca^{2+} + 6HPO_4^{2-} + 2H_2O \rightleftharpoons$$
$$Ca_{10}(OH)_2(PO_4)_6 + 8H^+$$

Figure 2. The model for proteolipid-dependent hydroxyapatite formation showing two proteolipids known to be involved in ion transport.

formation in both bacteria and vertebrate mineralization.

Aronson et al. (35) have shown that internal H^+ independent of its role as a substrate for exchange with external Na^+, has an important role as an allosteric activator of the Na^+-H^+ exchange system. Allosteric behavior with respect to internal H^+ is a property that would enhance the ability of plasma membrane Na^+-H^+ exchangers to extrude intracellular acid loads and thereby contribute to the regulation of intracellular pH. Because $[Na^+]_o/[Na^+]_i$ exceeds $[H^+]_o/[H^+]_i$ in most all conditions, the net driving force acting on these membrane ion exchangers actually enhances their ability to regulate intracellular pH against intracellular acid loads, possibly, such as would exist with HA formation. It is most important to realize that demonstration of these elegant pH regulating exchangers has been exclusively in plasma membrane vesicles, independent of direct cell regulation or ATP synthesis. This is exactly the situation that would exist in vertebrate matrix vesicle mediated mineralization and in B. matruchotii which calcifies its intracellular compartment only after the cell is in quienscience or death phase.

REFERENCES

1. Boyan, B., Knight, J., Landis, W., Dereszewski, G. and Zeagler, J. (1984) Scanning Electron Microscopy, 4, 1793-1800.
2. Boyan, B. (1985) in The Chemistry and Biology of Mineralized Tissues, (ed. Butler, W.) pp 126-131, EBSCO Media, Birmingham, Alabama, USA.
3. Bernard, G. (1972) J. Ultrastruct, 41, 1-17.
4. Gay, C.V. (1977) Calcif. Tiss. Res., 23, 215-223.
5. Anderson, G. and Boyan-Salyers, B. (1981) J. Dent. Res., 60A, 455.
6. Anderson, H.C. (1969) J. Cell Biol., 41, 59-72.
7. Boskey, A.L., Timchak, D.M., Lane, J.M. and Posner, A.S. (1980) Proc. Soc. Exp. Biol. Med., 165, 368-373.
8. Kim, K. (1976) Fed Proc., 35, 156-162.
9. Ennever, J., Vogel, J.J., Boyan-Salyers, B. and Riggan, L.J. (1978) J. Dent. Res., 58, 619-623.
10. Boyan-Salyers, B. and Boskey, A. (1980) Calcif. Tissue Int., 30, 167-174.
11. Boyan, B. and Boskey, A. (1984) Calcif. Tissues Int., 36, 214-218.
12. Ennever, J., Vogel, J.J. and Brown, L.R. (1972) J. Dent. Res., 51, 1483-1486.
13. Wadkins, C.L., Luben, R., Thomas, M. and Humphreys, R. (1974) Clin. Orthop., 99, 246-266.
14. Mandel, I.D. and Levy, B.M. (1957) Oral Surg., 10, 874-884.
15. Lear, J.D., Wasserman, Z.R. and DeGrado, W.F. (1988) Sci., 240, 1177-1181.
16. Benga, G. and Holmes, R.P. (1984) Prog. Biophys. Molec. Biol., 43, 195-257.
17. Saunders, V.A. (1978) Nature, 276, 116-117.
18. Korenbrot, J.I. (1977) Ann. Rev. Physiol. 39, 19-49.
19. Folch, J. and Lees, M. (1951) J. Biol. Chem., 191, 807-817.
20. Folchi-Pi, J. and Stoffyn, P. (1972) Ann New York Acad. Sci., 195, 86-107.
21. Boyan, B.D. and Clement-Cormier Y. (1984) in Membranes, Detergents, and Receptor Solubilization, pp 47-63.
22. Schlesinger, M. (1981) Ann Rev. Biochem., 50, 193-206.
23. Boggs, J.M., Wood, D.D. and Moscarello, M.A. (1981) Biochemistry, 20, 1065-1073.
24. Fischer, I. and Sapirstein, V.S. (1986) J. Neurochem., 47, 232-238.
25. Swain, L.D. and Boyan, B.D. (1988) J. Dent. Res., 67, 526-530.
26. Fillingame, R.H. (1980) Ann. Rev. Biochem., 49, 1079-1113.
27. Lind, C., Hojeberg, B. and Khorana, H.G. (1981) J. Biol. Chem., 256, 8298-8305.
28. Swain, L., Renthal, R. and Boyan, B. (1988) J. Dent. Res., in press.
29. Boyan, B., Schwartz, Z., Swain., L., Carnes, D. and Zislis, T. (1988) Bone, 9, 185-194.

30. Klionsky, D.J., Brusilow, W.S. and Simoni, R.D. (1983) J. Biol. Chem., 258, 10136-10143.
31. Hoppe, J., Gatti, D., Weber, H. and Sebald, W. (1986) Eur. J. Biochem,. 155, 259-264.
32. Brey, R.N. and Rosen, B.P. (1979) J. Biol. Chem., 254, 1957-1963.
33. Tsujibo, H. and Rosen, B.P. (1983) J. Bacteriol., 154, 854-858.
34. Abudkar, S.V., Zlotnick, G. W. and Rosen, B.P. (1984) J. Biol. Chem., 249, 6142-6146.
35. Aronson, P.S., Nee, J. and Suhm, M.A. (1982) Nature, 299, 161-163.
36. Aronson, P.S. (1985) Ann. N.Y. Acad. Sci., 456, 220-228.

SESSION II.
Calculus in Relation to
Plaque and Bacteria

A.A.Scheie

The role of plaque in dental calculus formation: a review

Faculty of Dentistry, University of Oslo, Geitmyrsveien 71, Oslo 4, Norway

ABSTRACT

Presence of bacteria is a characteristic of dental calculus, and available data suggest a complex role played by plaque bacteria in dental calculus formation. Bacteria may, actively or passively, promote mineralization by a nucleating function; they may increase the degree of supersaturation of the surrounding solution, thus making it more unstable; or they may distroy mineralization inhibitors. It is the intention of the present paper to review available literature related to these aspects.

Recent Advances in the Study of
Dental Calculus

INTRODUCTION

Dental calculus formation results from precipitation of calcium-
phosphate salts within the organic matrix provided by bacterial
plaque. At an early stage, small calcium-phosphate crystals may
be seen in the intermicrobial matrix. The matrix may gradually
become entirely calcified (1), leaving tubular holes corresponding to the shape of the bacteria (2,3). At a later
stage, the bacterial cytoplasm may also become mineralized (2).

Dental calculus consists of 70-80% inorganic, mainly crystalline salts (4). The organic composition consists of proteins,
carbohydrates, and a minor lipid fraction (5). It is comparable
to that of dental plaque (6).

Calculus deposition is preceded by plaque formation (1), and
the surfaces of supra- and subgingival dental calculus are
always covered by dental plaque (3,7-9). It is likely to assume
that these plaques, in some way, are involved in the calcification process; though presence of bacteria, reportedly, is not a
necessary condition for calculus formation. Dental calculus have
been found to form also in the absence of a microbial flora, as
observed on the teeth of germfree rats (10). However, the human
oral cavity is never free of bacteria, and it is generally
believed that bacteria play an important role in determining the
nature and the extent of dental calculus formation.

It is the intention of this paper to review available
literature on the role played by dental plaque bacteria in the
formation of dental calculus.

EPIDEMIOLOGIC DATA

Several epidemiologic surveys have tried to elucidate the
relationship between plaque and calculus formation by correlat-
ing index values of soft and mineralized tooth deposits (7).
The results from these studies are inconsistent. One reason may
be the problem of quantitating the soft and the mineralized
deposits with the available indices. Another point is that
plaque and calculus formation are dynamic processes, which can
hardly be reflected by cross-sectional data.

Though reducing the plaque load on the teeth may result in
reduced amounts of calculus, the great individual variation in
calculus formation cannot be explained by variations in plaque
formation rates (7).

STRUCTURAL STUDIES

The question has been raised whether specific bacteria, or
specific metabolic activities of certain bacteria may be linked
to calculus formation.

Structural studies of supragingival calculus and of calculus-
associated plaque indicate a dominance of filamentous organisms
(3). In plaque related to subgingival calculus, microorganisms
of varying morphology may be found (3).

MICROBIOLOGIC STUDIES

A wide range of bacterial strains have been identified in calculus and in calculus-associated plaques, by microbial cultivation (11). Qualitative analyses showed that organisms which are normally present in mature plaque, are also present in calculus (11). Up to twenty-two named isolates could be identified in one single sample (11). No particular organism seemed to dominate, neither was any particular organism consistently isolated from calculus samples (11). The findings were interpreted to indicate that calcification of plaque should be regarded as a consequence of specific activities of several microbial species, or as the result of a more general environmental change (11).

In theory, bacteria can, actively or passively, promote mineralization in several ways, 1: They may nucleate or seed the calcium and phosphate supersaturated solution surrounding them, 2: They may increase the degree of supersaturation of the surrounding solution by increasing the local pH, and thus make the solution more unstable, 3: They may destroy mineralization inhibitors present in saliva.

NUCLEATION

A number of oral bacteria become calcified upon in vitro growth in calcium-supplemented broths, or upon incubation in synthetic calcifying media (12-16). Both live, degenerated, and formalin-killed cells may become calcified (13,15,17,18). Several investigators have found that degenerated bacteria tend to calcify at an accellerated rate (13, 18, 19). In contrast, Streckfuss et al. (16) noted that rapidly growing streptococcal strains calcified more rapidly than slowly growing strains.

Gram-negative species have often been found to calcify more frequently, and more readily than Gram-positive bacteria (15). One exception seems to be the readily calcifiable Gram-positive Bacterionema matruchotii, which was mineralized in less than three days (15).

The nature of the calcium-phosphate precipitate may depend on the state of the bacteria. In vitro, octacalcium phosphate and brushite were more frequently deposited from formalin-killed cells than from viable cells (17).

Apparently the bacterial strain may also play a role in determining the mineralization pattern. Intracellular mineralization occurred in one B. matruchotii strain, whereas extracellular mineralization was present in another strain of the same species (13).

Nature of the nucleating factors
The nature of the nucleating factors derived from plaque bacteria has been subjected to extensive research. Loci of initial mineralization in virtually all vertebrate calcified tissues are membrane bound vesicles (20). It has been speculated that such vesicles are involved in mineralization of dental plaque.

Lie and Selvig (13) noted that the extracellular calcification by B. matruchotii was associated with trilaminar vesicles released by the cells. Several Gram-negative strains, among those, suspected periodontal pathogens, are also known to release membrane-derived micro-vesicles into the culture supernatant during growth (21,22).

In vitro studies have shown that isolated membrane-structures from the readily calcifiable B. matruchotii may function as nucleating agents able to precipitate hydroxyapatite and amorphous calcium-phosphate in vitro (23, 24). Closer analyses indicate that proteolipid complexes of the bacterial membrane are responsible for the nucleating action (25). The proteolipid complex includes an acidic phospholipid associated with a nonpolar protein. The acidic phospholipids have strong affinity for proteins, and this proteolipid complex has ability to bind calcium.

Vogel and coworkers (26) proposed a model illustrating the membrane facilitated calcification by proteolipid complexes. According to this model, the acidic phospholipids are stabilized by membrane bound proteins, and packed in such a way that negatively charged groups are located at opposite ends of the polar head of the molecule. These negatively charged groups bind calcium, readily. The calcium binding thus occurs as a two point electrostatic attachment between opposite ends of adjacent phospholipid molecules. The initial binding of calcium stabilizes the acidic phospholipid molecules in the membrane. A subsequent build-up of a calcium-phosphate unit structure occurs at each bound calcium. This so-called neoapatite structure, mimicing hydroxyapatite, is thought to provide a site for further crystal growth (26).

Ennever et al. (25) found that the proteolipids of B. matruchotii and the proteolipids which could be extracted from the calculus matrix, showed some general similarities; for instance the common occurrence of phosphoinositides and cardiolipin, and the degree of hydrophobicity (25). Phosphoinositides are acidic phospholipids common to all calcifiable proteolipids. Differences in the amino acid composition of the phospholipids were, however, present (25). Furthermore, the phospholipid composition of a calculus-extracted apatite nucleator was more complex than that extracted from B. matrushotii. The former contained both phosphatidyl ethanolamine and phosphatidyl serine whereas phosphatidyl serine was a minor component of the nucleator from B. matruchotii membranes (25). Both phosphatidyl ethanolamine and phosphatidyl serine are constituents of bacterial membranes (27). Phosphatidyl serine may also be transformed into phosphatidyl ethanolamine by bacterial enzymes (27). However, information about the presence of phosphatidyl serine in dental plaque is lacking. The reported differences between the calculus nucleator and the bacterial derived nucleator (25) may support the notion that the nucleators of dental calculus may originate, also from non-microbial sources, for instance from saliva. This assumption is supported by the findings of Slomiany et al. (28). Their findings also attest to the importance of lipids in the calculus formation; the lipid content of parotid saliva of high calculus formers was found to be higher than in saliva from subjects who formed little or no calculus (28).

Comparisons of the lipid fraction of calculus matrix and of parotid salivary gland stones may provide further information as to the source of the calculus nucleating factors. Dental calculus and parotid gland stones are similar concretions, except that the gland stones, usually, are formed in the absence of bacteria. The higher contribution of phosphatidyl ethanolamine in calculus (29) supports a bacterial origin, whereas the presence of cholesterol, both in dental calculus and in salivary gland stones, points to saliva as origin of the calculus nucleators (29, 30).

Importance of bacterial nucleation.
One point, which has been debated, is whether the mineralization starts inter- or intracellularly. The structural studies of in vivo formed calculus indicate that the mineralization of plaque starts between the bacteria, and that the bacteria often remain unmineralized (2). Friskopp (8) noted that the bacterial cell wall usually is the last structure to be mineralized. In contrast, in vitro mineralization seems to occur mainly intracellularly (17). Only very few cells induce extracellular calcification in vitro; exclusively intracellular calcification was found in the B. matruchotii strain studied by Sideway (17). This lack of consistency between in vitro and in vivo findings seems difficult to explain.

Most bacteria possess an ATP-dependent calcium extrusive system operating to maintain the internal calcium concentration lower than in the external medium (31,32). This prohibits intracellular calcification, unless the extrusive system is disturbed. As suggested by Sideway (18), it is possible that a certain degree of degeneration of the bacterial cells must take place before intracellular concentrations of calcium and phosphate reach the required levels. Experimental calcification of liposomes, artificial phospholipid vesicles, also supports this assumption. Reportedly, permeation of the liposomes by a ionophore was necessary in order to transport sufficient amounts of calcium to induce intravesicular mineralization (33).

As noted by Ooi and coworkers (34), many in vitro studies have applied calcium and phosphate concentrations in the calcifying solutions close to those at which spontaneous precipitation may occur. Under such conditions, even a very feable nucleator would cause precipitation. A relevant question may therefore be whether the in vitro nucleating ability found for a number of bacterial strains is relevant in vivo.

It is reported, however, that B. matruchotii undergoes calcification, in vitro at calcium and phosphate concentrations only slightly above those at which hydroxyapatite induces calcification (34).

BACTERIAL INFLUENCE ON THE DEGREE OF SATURATION

Enzyme histochemical studies of dental calculus indicate that mineralization of dental plaque should not merely be looked upon as a passive mineralization of dead bacteria, but also as a process promoted by enzymes in the covering dental plaque (9).

The minimum ion concentration required for nucleator-induced hydroxyapatite precipitation is strongly dependent on pH; elevated pH levels favor precipitation. It may therefore be presumed that the plaque, in periods, must be more alkaline than the surrounding fluids in order to induce precipitation of calcium-phosphate salts.

Local alkalinization may occur by bacterial release of ammonia through hydrolysis of urea or through metabolism of arginine. It is well documented that dental plaques possess ureolytic activity (35). The responsible bacteria are thought mainly to belong to the Actinomycetacea family (36). Some staphylococci and occational streptococci also show ureolytic activity (36).

The potential of plaque to increase precipitation of calcium-phosphate salts through ureolysis has been shown in vitro by the metabolism-dependent mineralization solution formulated by Pearce (37). The solution contains calcium, phosphate, urea and monofluorophosphate. The mineral content of plaque formed on intra-oral appliances increased by 40% after four days of treatment with the solution (37). Similar increases were later confirmed for plaque formed on natural teeth (38,39). Presumably, this increase was related to two metabolic properties of plaque bacteria. Hydrolysis of urea raises the pH by generating ammonia, whereas hydrolysis of monofluorophosphate increases the phosphate concentration (37). Both processes favor mineralization.

Interestingly, the arginine fermenting Streptococcus sanguis type I/II was found as the most prevalent streptococcus type in supra- and subgingival calculus samples, whereas S.sanguis type I is the most frequently isolated S.sanguis strain in unmineralized plaque (11). An S.sanguis type I/II strain isolated from dental calculus, failed, however, to calcify in an in vitro calcifying solution (15).

DESTRUCTION OF MINERALIZATION INHIBITORS

Pyrophosphatase
Pyrophosphate is known as a potent inhibitor of calcification processes, and the molecule is active in very low concentrations. Thus the ion concentration of calcium and phosphate necessary for precipitation of calcium-phosphate salts is greatly increased in the presence of pyrophosphate.

Pyrophosphate may be hydrolyzed by the enzyme pyrophosphatase, which greatly reduces the calculus inhibiting effect. Pyrophosphatase has been isolated from oral bacteria, among those Streptococcus mutans (40), and the calcifying microorganism B. matruchotii (41). The enzyme is not released during bacterial growth, and the pyrophosphatase activity was found to be highest in disrupted cells (40).

Dental plaque is known to exhibit pyrophosphatase activity, and in vivo calculus formation has been positively correlated with the level of acid pyrophosphatase activity in dental plaque

(42). It should be noted, however, that the pH of dental plaque probably does not favor the pyrophosphatase activity. Pyrophosphatase from S.mutans was found to have a pH optimum at 8.5 (40).

Pyrophosphatase activity has also been detected in saliva, and the level of activity was found to correlate with the calculus index (42). It is possible that pyrophosphatase contributes to the formation of dental calculus, but the origin of the pyrophosphatase cannot be clearly defined. Both saliva and plaque bacteria may be sources of pyrophosphatase activity.

Alkaline phosphatase also possesses pyrophosphatase activity, and the enzyme has been ascribed an important role in mineralization processes (43,44). This is ascribed to the pyrophosphatase activity, to the liberation of phosphate, and to a function of the enzyme in calcium transport (43,44).

Phosphatase activity is present in dental plaque bacteria (45,46), and in membrane-derived micro-vesicles of Gram-negative bacteria (22). Their role in calculus formation is, however, not verified.

Proteolysis
Statherine and acidic proline-rich proteins are potent inhibitors of calcium-phosphate precipitation. These proteins are constituents of saliva and have been ascribed roles as stabilizers of saliva calcium levels, through their binding of calcium. (47-49).

Proteolytic enzymes are able to break down proteins to peptides and amino acids. Proteolytic activity is present in dental plaque (50,51). In theory, the proteolytic activity of plaque may destroy mineralization inhibitors such as statherin, or acidic proline-rich proteins. Several proteolytic enzymes have been isolated from oral bacteria (52,53), also including suspected periodontal pathogens (46,54,55). The enzymes may be actively secreted by the bacterial cells during growth, or they may be released during autolysis. Extracellular membrane-derived micro-vesicles released from a number of Gram-negative bacterial species also possess proteolytic activity (22).

Plaque protease activity, reportedly, correlates with calculus formation (56).

A protease from Bacteroides Loescheii was shown to abolish the inhibition of calcium-phosphate precipitation by parotid saliva (57). The authors ascribed this to the hydrolysis of calcium-phosphate precipitation inhibitors.

The exact manner by which plaque proteases might operate in vivo remains, however, to be established.

SUMMARY

As outlined, available data suggest that plaque bacteria may play a complex role in the formation of dental calculus. However, since much data are derived from in vitro experiments, the relevance in vivo, and the relative importance of the various factors are difficult to assess. Definite conclusions are therefore difficult to draw.

Presumably, the calculus formed in vivo results from combined actions of various bacterial activities, by various bacterial species, under certain environmental conditions. The bacterial factors probably operate in combination with other host factors.

REFERENCES

1. Mislowsky, W.J. and Mazzella, W. (1974) J. Periodontol., 45, 822-829.
2. Shirato, M., Kamishikiryo, K., Itoh, A., Kado, H., Maeda, Y., Sekiguchi, T., Fekui, K. and Takezawa, T. (1981) J. Nihon Univ., 23, 179-187.
3. Friskopp, J. and Hammarström, L. (1980) J. Periodontol., 51, 553-562.
4. Leung, S.W. and Jensen, A.T. (1958) Int. Dent. J., 8, 613-626.
5. Mandel, I.D. (1963) Periodontics, 1, 43-52.
6. Silverman, G. and Kleinberg, I. (1967) Arch. Oral Biol., 12, 1387-1405.
7. Schroeder, H.E. (1969) Formation and inhibition of dental calculus. Hans Hubert publishers, Berne.
8. Friskopp, J. (1982) J. Periodontol., 54, 542-550.
9. Friskopp, J. and Hammarström, L. (1982) Acta Odont. Scand., 40, 459-466.
10. Fitzgerald, R.J. and McDaniel, E.G. (1960) Arch. Oral Biol., 2, 239-340.
11. Sidaway, D.A. (1978) J. Periodontal Res., 13, 349-359.
12. Ennever, J. (1968) J. Periodontol., 31, 304-307.
13. Lie, T. and Selvig, K.A. (1974) Scand. J. Dent. Res., 82, 8-18.
14. Ennever, J., Vogel, J.J. and Brown, L.R., jr. (1972) J. Dent. Res., 51, 1483-1486.
15. Sidaway, D.A. (1978) J. Periodontal Res., 13, 360-366.
16. Streckfuss, J.L., Smith, W.N., Brown, L.R. and Campbell, M.M. (1974) J. Bacteriol., 120, 502-506.
17. Sidaway, D.A. (1979) J. Periodontal Res., 14, 167-172.
18. Sidaway, D.A. (1980) J. Periodontal Res., 15, 240-254.
19. Rizzo, A.A., Martin, G.R., Scott, D.B. and Mergenhagen, S.E. (1962) Science, 135, 439-441.
20. Anderson H.C. (1976) Fed. Proc., 35, 105-108.
21. Hammond, B.F. and Stevens, R.H. (1982) in Host-parasite interactions in periodontal diseases, (eds. Genco, R.J. and Mergenhagen, S.) pp 46-61. Am. Soc. Microbiol., Washington DC.
22. Grenier, D. and Mayrand, D. (1987) Infect. Immunity, 55, 111-117.
23. Vogel, J.J. and Smith, W.N. (1976) J. Dent. Res., 55, 1080-1083.
24. Ennever, J., Vogel, J.J., Rider, L.J. and Boyan-Salyers, B. (1976) Proc. Soc. Exp. Biol. Med., 152, 147-150.
25. Ennever, J., Vogel, J.J., Boyan-Salyers, B. and Riggan, L.J. (1979) J. Dent. Res., 58, 619-623.
26. Vogel, J.J., Boyan-Salyers, B. and Campbell, M.M. (1978) Metab. Bone Dis. & Rel. Res., 1, 149-153.
27. Kates, M. (1964) Adv. Lipid Res., 2, 17-90
28. Slomiany, A., Slomiany, B.L. and Mandel, I.D. (1981) Arch. Oral Biol., 26, 151-152.

29. Slomiany, B.L., Murty, V.L.N., Aono, M., Sarosiek, J., Slomiany, A. and Mandel, I.D. (1983) J. Dent. Res., 62, 862-865.
30. Slomiany, B.L., Murty, V.L.N., Aono, M., Slomiani, A. and Mandel, I.D. (1983) J. Dent. Res., 62, 866-869.
31. Rosen, B.P. and Kashket, E.R. (1978) in Bacterial Transport, (ed. Rosen, B.P.) pp 559-620. Marcel Decker, N.Y.
32. Rosen, B.P. (1982) in Membrane Transport of Calcium, (ed. Carafoli, E.) pp 187-216. Academic Press, N.Y.
33. Eanes, E.D., Hailer, A.W. and Costa, J.L. (1984) Calcif. Tissue Int., 36, 421-430.
34. Ooi, S.W., Smillie, A.C. and Kardos, T.B. (1980) Can. J. Microbiol., 27, 267-270.
35. Frostell, G. (1960) Acta Odont. Scand., 18, 29-65.
36. Gallagher, I.H.C., Pearce, E.I.F. and Hancock, E.M. (1984) J. Dent. Res., 63, 1037-1039.
37. Pearce, E.I.F. (1981) Front. Oral Physiol., 3, 108-124.
38. Pearce, E.I.F., Schamschula, R.G. and Cooper, M.H. (1983) J. Dent. Res., 62, 818-820.
39. Schamschula, R.G., Pearce, E.I.F., Un, P.S.H. and Cooper, M.H. (1985) J. Dent. Res., 64, 454-456.
40. Wöltgens, J.H.M., Bervoets, T.J.M. and de Vries, W. (1977) J. Periodontal Res., 12, 462-466.
41. Pellat, B.P. and Grand, M. (1986) J. Biol. Buccale, 14, 223-228.
42. Bercy, P. and Vreven, J. (1979) J. Biol. Buccale, 7, 31-36.
43. McComg,R.B., Bowers, G.N., jr. and Posen, S. (1979) Alkaline phosphatase, pp 872-883. Plenum Press, N.Y.
44. Ten Cate, A.R. (1980) Oral Histology, Development, Structure, and Function. pp 94-109. C.V. Mosby Company, Saint Louis.
45. Bowen, W.H. (1961) J. Dent. Res., 40, 571-577.
46. Greeman, J. and Melville, T.H. (1978) Arch. Oral Biol., 23, 965-970.
47. Oppenheim, F.G., Hay, D.I. and Franzblau, C. (1971) Biochemistry, 10, 4233-4238.
48. Hay, D.I. and Oppenheim, F.G. (1974) Arch. Oral Biol. 19, 627-632.
49. Hay, D.I. (1973) Arch. Oral Biol., 18, 1531-1541.
50. Söder, P-Ö. and Frostell, G. (1966) Acta Odont. Scand., 24, 501-515.
51. Söder, P-Ö. (1972) J. Dent. Res. Suppl. no. 2, 51, 389-393.
52. Frostell, G., Nord, C-E. and Söder, P-Ö. (1973) Odont. Revy, 24, 27-38.
53. Sato, S., Koga, T. and Inoue, M. (1983) Arch. Oral Biol., 28, 211-216.
54. Fujimura, S. and Nakamura, T. (1981) Infect. Immunity, 33, 738-742.
55. Slots, J. and Genco, R.J. (1984) J. Dent. Res., 63, 412-421.
56. Morita, M. and Watanabe, T. (1986) J. Dent. Res., 65, 703-705.
57. Morishita, M., Tokumoto, K., Watanabe, T. and Iwamoto, Y. (1986) Arch. Oral Biol., 31, 555-557.

A.Tatevossian

Is plaque fluid composition related to a predilection for calculus formation?

18 Thornhill Road, Llanishen, Cardiff CF4 6PF, Wales, UK

ABSTRACT

Resting plaque fluid is highly supersaturated with respect to enamel mineral and is slightly so with respect to dicalcium phosphate dihydrate. Experiments were devised to test the hypothesis that plaque residue contains sources of calcium whose dissociation maintains the high concentration in plaque fluid.

Pooled 24h plaque from student volunteers was centrifuged at 5000 g following which plaque fluid was decanted. The residues were homogenised in 1-10 mmol/L calcium standard solutions, pH 6.5 at 25 °C for 30 min. The residues were then centrifuged, their supernates decanted for separate analysis, and then sequentially extracted with 0.2 mol/L acetate pH 7.5, 0.2 mol/L acetate pH 5.0, 0.1 mol/L NaOH. The supernates and extracts were assayed for ionized calcium and total calcium.

Plaque residue uptake of calcium from the standard solutions was linearly correlated with the concentration used. This calcium was extracted by 0.2 mol/L acetate pH 7.5 and 5.0, and a smaller fraction was released only after incubation with 0.1 mol/L NaOH in the cold.

From the results it is concluded that a dissociable reservoir of calcium is present in plaque residue and it is suggested that this reservoir may buffer perturbations in plaque fluid calcium concentrations. It follows from this that resting plaque fluid composition is related to a predilection for calculus formation.

Recent Advances in the Study of
Dental Calculus

INTRODUCTION

Supragingival calculus formation has been found to be associated with an intracellular calcification in dental plaque bacteria (1,2). This process can be demonstrated in vitro with a wide selection of permanent and transient oral bacterial species (3). The mineral phases which can be identified in calculus have been the subject of considerable study (4-8) and their formation and transformations are reviewed by several authors in this workshop (9,10). However, there are few studies of dental plaque composition in relation to calculus formation.

Early in the study of plaque composition, it was noted that plaque collected from sites which are prone to calculus tends to have higher total concentrations of calcium and phosphate (11). These and other inorganic components were examined in 3-day plaque among groups classified as 'heavy' and 'light' calculus formers by Mandel (12). He confirmed the findings of Dawes and Jenkins concerning the sites with higher calcium and phosphate concentrations being more prone to calculus formation. The 'heavy' calculus formers had a higher concentration of calcium and phosphorus in the plaque fraction extractable in water, presumably reflecting partly its plaque fluid component. There did not seem to be major differences in the insoluble plaque fractions from the two groups.

Intraoral deposits collected for 24 h on mylar strips take up calcium and phosphate from a mineralizing solution (13). The plaque samples from subjects classified as 'heavy calculus formers' showed greater uptake than those from 'light calculus formers', using the weight of calculus removed at three successive scalings as a criterion. A study of the nature of calcium uptake by dental plaque showed that it reached near saturation within 10 minutes, was independent of the viability of the biomass but was affected by ionic strength, calcium ion concentration and inoculum density (14). An increase in ionic strength and a reduction in pH caused a rapid release of bound calcium.

We recently determined in plaque fluid the ion products for calcium phosphate species which are likely to be found at the enamel-plaque-calculus interfaces (15). We found that in overnight-fasted dental plaque, the aqueous millieu is ~8 orders of magnitude above saturation with respect to enamel mineral, ~1 order of magnitude above saturation for octacalcium phosphate and is even slightly above saturation with respect to dicalcium phosphate dihydrate. Since homologous saliva concentrations of the relevant ions are below those in plaque fluid, a source for such ions within dental plaque residue is hypothesized. In the case of calcium ion, earlier studies (14) are consistent with this hypothesis. This hypothesis was tested further by the experiments reported here, whose aim was to investigate some of the conditions for the dissociation of calcium found in plaque residue and to interpret the data in relation to mineralisation leading to calculus being formed.

MATERIALS AND METHODS

Pooled 24h dental plaque was collected from student volunteers at least 1h after their mid-morning snack using specially made perspex spatulae. Before collection, subjects were asked to swallow so as to minimise oral salivary content. Plaque was taken from all available tooth surfaces except the lingual aspects of lower incisors and canines, immediately plunged into liquid nitrogen and pooled within 1-1.5h after collection of the first student's sample.

Pooling of plaque was carried out in a cold room at $5^{\circ}C$, after which the samples were centrifuged at 5000 g for 15 min at $5^{\circ}C$. The supernatant plaque fluid was decanted. The residue was divided into aliquots (10 mg wet weight). Each aliquot was vortexed in 1.0 mL of calcium chloride standard in 0.2 mol/L triethanolamine buffer pH 6.5. This buffer was previously shown to be satisfactory for calcium ion determinations (16) and appeared not to bind any calcium in solution. The vortexed samples were left for 30 min at $25^{\circ}C$, then centrifuged at 5000 g for 10 min at $25^{\circ}C$. Some of the samples were continuously monitored at $25^{\circ}C$ for the kinetics of calcium uptake using the calcium electrode assembly previously described (16).

The supernatants were decanted after centrifugation. The residues were inverted to drain undecanted supernatant and then vortexed in 1.0 mL 0.2 mol/L acetate buffer pH 7.5 at $25^{\circ}C$ for 5 min before centrifugation under the conditions described above. The supernatants were decanted, the residues drained as described and then vortexed in 1.0 mL 0.2 mol/L acetate buffer pH 5.0 at $25^{\circ}C$ for 5 min. Following a further round of centrifugation and decanting, the residues were then vortexed in 0.5 mL of 0.1 mol/L NaoH and left to stand 1h at $25^{\circ}C$. The supernatant from this extraction was also separated by centrifugation. All supernates were recentrifuged before sampling for analysis.

Calcium ion concentrations were determined using an inverted mode electrode (16). The calcium exchanger, bis(di[p-1,1,3,3-tetramethylbutylphenyl] phosphate) and a trialkyl phosphate mediator were incorporated into a poly(vinyl chloride) matrix. Total calcium concentrations were determined on a Pye Model 290 atomic absorption spectrophotometer using lanthanum chloride to suppress phosphate interference. There was good agreement between the results from these independent assays and the results were combined for presentation.

The mass action law was applied to the data on calcium binding using the following assumptions: 1) that the valencies of the binding components could be averaged out as divalent anions 2) that the molecular weights of binding components could be averaged out at 10^5 daltons. Although these are recognized as arbitrary criteria, on average they should be expected to apply uniformly to all samples. Similarly, since the calcium binding rate showed saturation kinetics, a double reciprocal plot derived from the Michaelis-Menten equation was used to quantify the kinetics of the process.

RESULTS

There was a significant variation in the uptake of calcium by the samples of dental plaque residue, some showing little or no uptake (Table I). The uptake of calcium showed a linear regression with the concentration of calcium used in the standard solutions (Table I).

Table I. CALCIUM UPTAKE BY DENTAL PLAQUE FROM CALCIUM
SOLUTIONS (pH 6.5) AT CONSTANT IONIC STRENGTH

| | Original calcium concentration (mmol/L) | | | | |
	1.0	2.5	5.0	7.5	10.0
Final Ca (mmol/L)					
Mean	0.78	2.13	4.46	6.48	9.08
Range	1.03-0.54	2.6-1.75	5.3-3.8	7.6-5.47	9.8-8.28

Correlation coefficient, r = 0.99, n=50, p<0.001

This uptake was designated as bound calcium and its
dissociation, as evaluated by sequential extraction, was
apparently greater in acetate buffer pH 7.5 (Table II).

Table II. DISSOCIATION OF CALCIUM BOUND BY DENTAL PLAQUE
(nmol Ca/mg wet weight)

| | Uptake from calcium solution at (mmol/L) | | | | |
	1.0	2.5	5.0	7.5	10.0
BOUND Ca	21.6	37.5	53.7	101.8	91.6
Ca REMOVED BY EXTRACTION IN					
Acetate pH7.5	18.4	25.8	33.2	47.8	65.0
Acetate pH5	15.6	10.6	10.8	16.5	19.7
NaOH	10.2	7.9	7.3	7.5	6.9
Acetate pH5 +NaOH	25.8	18.5	18.1	24.1	27.4
TOTAL Ca FROM ALL FRACTIONS	44.2	43.5	51.3	71.9	92.4

However, this first extract was contaminated with the calcium
standard used for uptake which would remain trapped within the
biomass. This fraction, together with that extracted by acetate
at pH 5.0 accounted for the major part of the calcium uptake.
The sum of the calcium found in the extracts was in close
agreement with the amount bound, as assessed by the change in

Table III. MASS ACTION CALCULATION OF CALCIUM BINDING DATA

$$K = \frac{[Ca^{2+}][B^{2-}]}{[CaB]}$$

| | Uptake from Calcium solution at (mmol/L) | | | | |
	1.0	2.5	5.0	7.5	10.0
K (mmol/L)	0.76	1.75	3.92	5.98	9.23

calcium concentration of the standard solutions after mixing with dental plaque (Table II). However, a smaller but significant fraction of the bound calcium was released only after digestion in NaOH (Table II).

Using the mean uptake data from calcium standard solutions at different concentrations, the mass action law was evaluated for the binding of calcium (Table III). The absence of a concentration equilibrium constant is notable.

Table IV. KINETICS OF CALCIUM BINDING IN DENTAL PLAQUE

r	m	b	p<	V_{app} (nmol/mg/min)	K_{app} (mmol/L)
0.99	0.05	0.015	0.001	67.3	3.7

An evaluation of the kinetics of the process, from which a double reciprocal plot of the rate of calcium uptake as a function of the calcium concentration in the standard solutions was obtained. The kinetics conformed to the characteristics of a saturable process and the apparent maximal velocity for the uptake and the apparent Km for the process were obtained from the graph (Table IV).

DISCUSSION

The uptake by 24h dental plaque of calcium and phosphate from metastable mineralising solutions was reported by Eilberg et al. (13). However, calcium uptake can be demonstrated in stable buffered solutions and some of the attributes of this uptake were reported at a previous meeting of this group (14).

The uptake of calcium from standard calcium solutions increased with concentration in the range 1-10 mmol/L (Table I). At each concentration some samples showed little or no uptake. Eilberg et al. (13) noted that 24h human plaque samples from 'light' calculus formers showed little or no uptake of calcium or phosphate while plaque from 'heavy' calculus formers showed high levels of uptake. In the present study no data was available on the rate of formation of calculus by the plaque donors. However, the variability and heterogeneity in the composition of pooled samples of dental plaque may adequately account for the scatter in the results.

The change in calcium concentration of the standard solutions 30 min after vortexing with dental plaque could not have been ascribed to precipitation of calcium phosphates since, at pH 6.5 and hundred-fold volumetric dilution, soluble plaque calcium and phosphate concentrations would be far below the solubility products of the calcium phosphates. This validates the designation of such uptake as bound calcium.

The calcium binding was apparently not under equilibrium conditions, as judged by the variable binding constants which were calculated (Table III). This could not be ascribed to the kinetic nature of the binding since earlier data, confirmed in the present series, indicated that the calcium binding, as defined above, reached a plateau within 10 min; a contact time of 30 min with the standard calcium solutions was more than 3x longer than was necessary to obtain stability in the reaction mixtures. The same conclusion can be drawn from the results of the kinetic plot (Table IV), which indicated an apparent Km of 3.7 mmol/L and apparent Vmax of 67 nmol/ mg/min for the calcium binding. Despite the tenuous nature of the assumptions made for these calculations, the increase in binding constant with increasing concentrations of calcium standard solutions implies the recruitment of binding sites of different affinities and valencies within the biomass. Some of the plaque constituents which can be demonstrated to show such properties include bacterial teichoic acids and other fixed anionic sites (17,18), calcium-binding proteins (19-21) and calcium-precipitable protein (22-24) originating from saliva, and intracellular compartments of calcium in plaque bacteria (1,3,25).

The bound calcium was compartmentalised, as indicated by the serial extraction in acetate pH 7.5, then pH 5.0, then in NaOH. The fraction of bound calcium extracted in acetate pH 5.0 and in NaOH did not increase significantly with increasing calcium concentration (Table II). The bound calcium in these two fractions accounted for a relatively constant amount of around 24 nmol Ca/mg wet weight while the fraction extracted by acetate pH 7.5 increased significantly with the concentration of calcium standard. This increase accounted for most of the uptake at >2.5 mmol/L Ca standard (Table II). The sum of Ca found in the three fractions agreed well with the bound calcium calculated from the change in calcium concentration of the standard solutions after standing 30 min with the plaque samples. Two possible interpretations of this data are as follows.

At face value, the results suggest that the largest component of bound calcium is loosely bound, that this compartment shows a concentration-dependent binding and that its bound calcium is readily extracted by acetate pH 7.5. On this premis, the fractions extracted by acetate pH 5.0 and by NaOH, which showed no significant increase with increasing calcium standard concentrations, may represent calcium native to the plaque samples at the time of collection. However, most of the plaque calcium has been reported to be extractable by acetate pH 5.0 (11) and little or none at pH 7.5. This finding was confirmed more recently (14).

On the other hand, the large calcium component in acetate pH 7.5 may be explained if it is assumed that the extract contained residual calcium from the standard solutions which could not be drained completely and which may also have remained trapped within the centrifuged plaque biomass. It can be calculated that a contamination with 6.5-18 microlitres of 10-1 mmol/L calcium standard solution resectively could account for the entire calcium found in these fractions. Since this represents <2% of the volume of calcium standard solutions which were used, it is a probable error factor which could not be eliminated from the methodology in obtaining the first extract. If the data in the remaining fractions is reviewed with this in mind, the bulk of the bound calcium was probably in the compartment which was extracted by 0.2 mol/L acetate pH 5.0, with a significant but smaller fraction being released only after incubation with 0.1 mol/L NaOH. Probably none of the extractions were 100% efficient in removing calcium, suggesting that the agreement between uptake and total bound calcium data is partly fortuitous.

Although the compartmentalisation of bound plaque calcium remains somewhat obscure, the results show that a dissociable reservoir of calcium is demonstrable in plaque residue. The rapid kinetics of the uptake suggests that the compartments are in the matrix or extracellular phase since calcium ingress into the bacterial cells is slow and its extrusion by active transport is well documented (26). Taken in conjunction with earlier findings on plaque fluid composition during the time-course of a Stephan curve, it may be expected that the binding capacity in the residue buffers plaque fluid calcium concentration when it is perturbed, thus maitaining the supersaturation in plaque fluid with respect to calcium phosphate species. This would lead to a constant and steep gradient favouring calcium ingress into cells and cells with compromised calcium transport could accumulate high intracellular calcium concentrations. This implies that resting plaque fluid composition is related to a predilection for calculus formation. This is consistent with literature data that early mineralization can be shown in dental plaque accumulations of >4days (27).

ACKNOWLEDGEMENT
    This study was supported by a Project Grant from the Medical Research Council.

REFERENCES

1. Ennever, J. and Creamer, H. (1967) Calc. Tiss. Res., 1,87-93.
2. Mandel, I.D. and Gaffar, A. (1986) J. Clin. Perio., 13, 249-257.
3. Streckfuss, J.L., Smith, W.N., Brown, L.R. and Campbell, M.M.

(1979) J. dent. Res., 58, 1916-1917.
4. Rowles, S.L. (1964) Proc. 1st Eur. Bone and Tooth Symp., Pergamon Press, Oxford, UK.
5. Saxton, C.A. (1968) Archs oral Biol., 13, 243-246.
6. Newesely, H. (1968) Caries Res., 2, 19-26.
7. Le Geros, R.Z. (1974) J. dent. Res., 53, 45-50.
8. Killian, W.F. and Ennever J. (1975) J. dent. Res., 54, 185.
9. Driessens, F.C.M. and Verbeeck, R.M.H. (1989), This volume.
10. White, D.J., Bowman W.D. and Nancollas, G.H. (1989), This volume.
11. Dawes, C. and Jenkins, G.N. (1962) Archs oral Biol., 7, 161-172.
12. Mandel, I.D. (1974) J. Periodont. Res., 9, 10-17.
13. Eilberg, R.G., Judy, K., Iovino, E., Kornfeld, P., Phelan, J. and Ellison, R. (1973) J. dent. Res., 52, 45-48.
14. Tatevossian, A. (1981) in Tooth Surface Interactions and Preventive Dentistry, (eds. Rolla, G., Sonju, T., Embery, G.) pp 105-112. IRL Press, London, UK.
15. Carey, C., Gregory, T., Rupp, W., Tatevossian, A. and Vogel, G.L. (1986) in Factors Relating to the Demineralisation and Remineralisation of the Teeth, (ed. Leach, S.A.) pp 163-173. IRL Press, Oxford, UK.
16. Tatevossian, A. (1987) Archs oral Biol., 32, 201-205.
17. Knox, K.W. and Wicken, A.J. (1973) Bacteriol. Rev., 37, 215-257.
18. Rölla, G. and Bowen, W.H. (1977) Scand. J. dent. Res., 85, 149-151.
19. Grön, P. and Hay, D.I. (1975) J. dent. Res., 48, Suppl., 799-805
20. Hay, D.I. and Schlesinger, D.H. (1977) in Calcium-binding Proteins and Calcium Function, (eds. Wasserman, R.A., Corradino, R.A., Carafoli, E., Kretsinger, P.H., MacLennan, D.H. and Siegel, F.L.) pp 401-408. North Holland, Oxford, UK.
21. Leach, S.A. (1980) in Dental Plaque and Surface Interactions in the Oral Cavity, (ed. Leach, S.A.) pp 159-183. IRL Press, London, UK.
22. Bettelheim, F.A. (1971) Biochim. Biophys. Acta, 236,702-705.
23. Boat,T.F., Wiesman, U.N. and Pallavacini, J.C. (1974) Pediat. Res., 8, 531-539.
24. Edgar, W.M., Jenkins, G.N. and Hillam, D. (1980) in Dental Plaque and Surface Interactions in the Oral Cavity, (ed. Leach, S.A.) pp 197-210. IRL Press, London, UK.
25. Sidaway, D.A. (1978) J. Periodont. Res., 13, 360-366.
26. Silver, S. (1977) in Microorganisms and Minerals (ed. Weinberg, E.D.) pp 50-103. Marcel Dekker Inc., New York, USA.
27. Schroeder, H.E. (1963) Helvet. odont. Acta, 7, 17-30.

T.Watanabe
M.Morita

Relation between calculus formation and the protease activity of saliva/plaque

Department of Preventive Dentistry, Okayama University Dental School, Shikata-cho, Okayama, 700 Japan

ABSTRACT

Here is suggested that the protease of dental plaque might promote calcification of dental calculus. The positive correlation between the calculus index and the salivary protease activity was demonstrated. The origin of the protease might be some microorganisms. Saliva contains the inhibitors of proteins to precipitate the calcium phosphate from the supersaturated solutions. The inhibitory activity was degenerated by the protease from <u>Bacteroides</u> <u>loescheii</u> from saliva. It is reasonable to consider that protease activity in dental plaque, rather than salivary protease activity, can be related to calculus formation. Correlation between dental calculus and plaque protease was confirmed. And it suggests that <u>Capnocytophaga</u> protease may contribute to dental calculus formation, especially mineralization of dental plaque.

INTRODUCTION

It is generally agreed that protease is an important factor in periodontal disease. On the other hand, proteases reduce the mineralization of plaques. Patients treated with pancreas powder showed a 60% inhibition of new dental calculus formation[1]. Demmers and Belting[2] demonstrated that proteolytic fungal enzyme Molsin(Aspergillo peptidase-A), and two trypsin preparations were 80 to 100 percent effective inhibitors[3].

In 1980, the salivary protease activities used hemoglobin as a substrate were demonstrated positive correlations with calculus index[4] and ages, not but DMFT index of dental caries, P-M-A index of gingivitis[5], the debris index and salivary protein[6] (Table 1). The results were confirmed as the protease of salivary supernatant was correlated well with the supragingival calculus index, and the protease of salivary sediment showed significant correlations with subgingival calculus index, which were analyzed by a first-order partial correlation coefficient[7].

ORIGIN OF THE SALIVARY PROTEASE RELATED TO CALCULUS

The correlation coefficients between the protease activities and the calculus index were positive and statistically significant, but not significant between the protease and the debris index. From the result, three possibilities were considered as follows; (a) the protease activity might catalyze a calcification of plaque, (b) cells related to calculus formation, i.e. microorganisms, leukocytes, epithelial cells, might release proteases, (c) protease activity might increase with gingival crevicular fluid owing to calculus deposition (Fig. 1).

Epithelial cells and protease in saliva
The origin of proteases in oral cavity were not clear, even though they had been considered to be leukocytes from the gingival exduate[8], dental plaque and microorganisms[9,10].

Table 1. CORRELATION COEFFICIENTS BETWEEN PROTEASE ACTIVITIES AND CLINICAL INDICES, AGE, AND CONCENTRATION OF SALIVARY PROTEIN[6]

| | Protease(pH5.5) | | Protease(pH8.5) | |
	unit/mL	S.A.	unit/mL	S.A.
DMFT index	-0.069	-0.130	-0.041	-0.112
P-M-A index	0.133	0.150	0.188	0.203
Debris index	-0.130	0.053	-0.150	-0.053
Calculus index	0.442**	0.405 **	0.449 **	0.411 **
Age	0.456**	0.322 *	0.496 **	0.384 *
Protein	0.484**	0.102	0.482 **	0.132

n=77
S.A.; Specific activity was expressed as units per mg of protein. * $p<0.01$, ** $p<0.001$

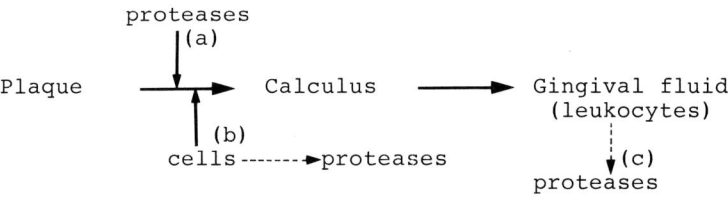

Figure 1. Three possibilities to explain the relationship between the proteases and calculus.

Table 2. PROTEASE ACTIVITY IN HUMAN SALIVA [11]

	Whole Saliva	Parotid Saliva	Submaxillary Saliva
Protease, pH5.5			
(unit/mL)	0.889±0.885	N.D.	0.004±0.009
(specific activity)	0.232±0.135	–	0.008±0.002
Protease, pH8.5			
(unit/mL)	0.905±0.790	N.D.	0.016±0.017
(Specific activity)	0.237±0.118	–	0.020±0.039

n=10, Mean±S.D.
N.D.: Not detectable.

The protease activities were measured in glandular saliva and whole saliva [11]. No protease activities was found in the ten parotid saliva samples. Six out of ten submaxillary saliva samples showed no activity, but the others showed very low activities comparing with that of the whole saliva. This suggests that glandular saliva is not the main source of salivary proteases(Table 2).

To eliminate the factors of crevicullar fluid, the saliva samples were collected from edentulous persons. Those people whose ages were almost the same as the edentulous group were selected from patients with marginal periodontitis or gingivitis as a control. The protease activities, number of leukocytes, that of epithelial cells and protein concentration were analyzed in saliva samples. The results were summerized in Table 3 [11]. The number of salivary leukocytes of the edentulous group was about one-half of that of the control group, but there was no statistical significance. The number of epithelial cells, the protease activities, and the protein concentrations between the both groups showed few differences. These results suggest that the main origin of the salivary protease is not leukocytes nor crevicular fluid(Fig. 1).

The correlation coefficients between the protease activities and number of epithelial cells were calculated before and after oral prophylaxis [11]. The positive significant correlation coefficients were observed between the number of epithelial cells and the protease activities at pH5.5 and pH8.5.

Table 3. PROTEASE ACTIVITIES IN SALIVA FROM EDENTULOUS PERSONS
AND PATIENTS WITH GINGIVITIS [11]

	Gingivitis (n=9)	Edentulous (n=10)	Student's t-test
Age	51.9 ±2.5	59.9 ±11.6	t= 1.9079
Leukocytes (X10^5/mL)	1.51 ±0.93	0.88 ± 0.59	1.6184
Epithelial cells(X10^5/mL)	4.02 ±4.50	4.31 ± 3.66	0.1417
Protein(mg/mL)	2.52 ±0.84	2.91 ± 0.64	1.0409
Protease, pH5.5			
(unit/mL)	0.515±0.401	0.703± 0.661	0.6883
(specific activity)	0.194±0.123	0.241± 0.221	0.5250
Protease, pH8.5			
(unit/mL)	0.806±0.754	0.824± 0.878	0.0426
(specific activity)	0.291±0.185	0.284± 0.308	0.0604

Mean ± S.D.

Table 4. THE COEFFICIENTS OF CORRELATION BETWEEN THE PROTEASES
AND LEUKOCYTES OR EPITHELIAL CELLS [11]

	Leukocytes		Epithelial cells	
	Oral prophylaxis		Oral prophylaxis	
	Before	After	Before	After
Protein (mg/mL)	r= 0.002	0.002	0.516**	0.273
Protease, pH5.5				
(unit/mL)	-0.034	-0.115	0.451**	0.533***
(Specific activity)	-0.001	-0.116	0.253	0.442**
Protease, pH8.5				
(unit/mL)	-0.061	-0.163	0.561***	0.788***
(Specific activity)	-0.024	-0.201	0.404*,+	0.743***,+
Calculus index	0.109	-0.096	-0.062	0.169

n=37
* $p < 0.02$, ** $p < 0.01$, *** $p < 0.001$
† Significant difference between the two coefficients: $p < 0.03$

There was a significant difference of the two coefficients of
epithelial cells at the specific activities of pH8.5 before
and after oral prophylaxis(Table 4). The number of leukocytes
had no correlation with the protease activities(Fig. 2). It
was confirmed that the leukocytes might not be a main origin
of the proteases in saliva.

It might be possible to consider from these results that a
main source of salivary protease is the epitheliall cells in
saliva [11]. On the other hand, some of the proteases in dental
plaque are known to be of bacterial origin[12]. A number of
streptococci are reported to elaborate extracellular proteases
in oral cavity [10]. Orban and Weinmann [13] described epithelial
cells in saliva were covered by microorganisms. There is also
a possibility that a significant positive correlation between
the protease activity and the epithelial cells contents depend
on microorganisms covering the epithelial cells.

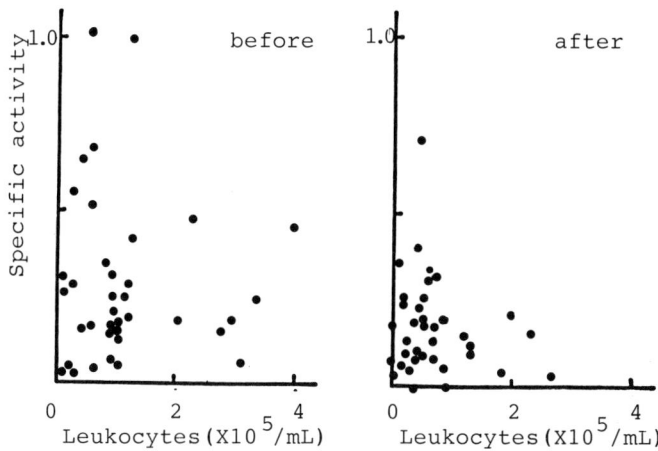

Figure 2. Correlation between the specific activity of protease and the number of leukocytes in saliva. Left, before oral prophylaxis; right, after oral prophylaxis [11].

Table 5. CORRELATION COEFFICIENTS OF THE PROTEASE WITH THE CALCULUS INDEX AND THE NUMBER OF EPITHELIAL CELLS [15]

Protease(pH7.0)	Calculus index	Epithelial cells
Whole saliva	r= 0.4662 **	r= 0.3530 *
Filtrate	0.4378 **	0.2473
Unfiltrate	0.2541	0.4521 **

n=37; * $p<0.05$, ** $p<0.01$

Table 6. SPECIFIC ACTIVITY OF THE PROTEASE IN THE DENTAL PLAQUE SUSPENSION [17]

Plaque suspension	Specific activity in the dental plaque	
	Non-calculus group	Calculus group
with T.G.C.	8.8 ± 7.2 (37) *	14.0 ± 8.5 (30) *
without T.G.C.	6.1 ± 3.7 (24)	6.3 ± 3.3 (18)

Mean ± S.D.,   ( ); number of the patients
T.G.C.; sodium thioglycollate
* Significant difference between the two groups by analysis of variance ($p<0.01$)

Protease activity related to calculus index.
    No significant correlations of the calculus index were demonstrated with the calcium concentration, the phosphate concentration, the number of leukocytes and the number of epithelial cells in saliva, while the calculus index correlated well with the protease activity in saliva[14] . The origin of the protease related to the calculus index was

69

thought to be different from the epithelial cells in saliva (Table 4). Tokumoto et al.[15] investigated the relationship between the calculus index and the whole saliva protease excluding the epithelial cells.

Precipitated whole saliva was suspended into 0.9% saline and was devided into the two fractions by the membrane filter (12μ). The unfiltrate material was recovered with 0.9% saline. The protease activity and the number of epithelial cells were analyzed in the three fractions; the precipitated whole saliva, the filtrate and the unfiltrate. The recovery of the epitherial cells was 27% in the filtrate and 52% in the unfiltrate; 79% in total. The protease activity was 52% in the filtrate and 23% in the unfiltrate. The results were summerized in Table 5. The protease activity in the filtrate correlated well with the calculus index, and the protease in the unfiltrated showed a significant correlation with the number of epithelial cells.

The results suggest that an origin of the protease related to calculus index is different from epithelial cells in saliva. Cellular composition of whole saliva is 86% of epithelial cells, 13.7% of leukocytes and 0.3% of microorganisms[16]. It would seem reasonable to assume that the origin of the protease related to the calculus index is microorganisms.

CALCULUS AND PROTEASE ACTIVITY IN DENTAL PLAQUE

Calculus formation results from mineralization of dental plaque, and dental calculus is usually covered with metabolically active plaque. Therefore, it is reasonable to consider that protease activity in dental plaque, rather than salivary protease activity, can be related to calculus formation.

Morita and Watanabe demonstrated that the protease activity in dental plaque from the patients with dental calculus is higher than those without calculus(Table 6)[17]. Plaque specimens from the calculus group showed significantly greater protease activity in the presence of 0.05% sodium thioglycollate than from the non-calculus group. No significant difference between the two groups was observed in the protease activity without sodium thioglycollate. These results suggest that the protease correlated calculus was a thiol protease, because sodium thioglycollate protects SH residues such as that in cystein.

Identification of the proteolytic microorganisms related to calculus formation was investigated[18]. One hundred and seventeen periodontal patients were devided into two groups; the calculus group and the non-calculus group. The plaque samples on the lingual surface of the lower anterior teeth with or without supragingival calculus were cultured on the selective mediums. The colony forming units ratios in the Bacteroides agar were significantly different between the two groups(Table 7)[18]. Bacteroides intermedius, Bacteroides melaninogenicus and Capnocytophaga species were observed more frequently in the calculus group than in the non-calculus group(Table 8). The Capnocytophaga species showed more than about ten times of the protease activity of the Bacteroides

Table 7.   COLONY FORMING UNITS RATIOS(%) OF SUPRAGINGIVAL
           PLAQUE USING THE SELECTIVE MEDIUMS [18]

Selective medium	Calculus group (n=29)	Non-calculus group (n=40)
MS agar	19.6 ± 13.4	26.0 ± 23.0
FM agar	1.0 ± 1.5	0.7 ± 1.3
VM agar	10.7 ± 11.7	11.8 ± 16.7
SL agar	1.3 ± 3.2	2.0 ± 6.5
Bact. agar	11.5 ± 11.5 *	5.0 ± 6.2 *
MSB agar	2.5 ± 3.5	4.5 ± 7.6

Mean ± S.D.
* Significant difference between the two groups by analysis of
  variance ($p<0.01$)

Table 8.   PREVALENCE OF THE BACTERIA ON THE SELECTIVE MEDIUM [18]

Bacterial species	Calculus group	Non-calculus group
Bacteroides	39/49 *	14/68 *
Capnocytophaga	38/49 †	24/68 †

*,† : Significant difference between the two groups by
      x^2-test($p<0.001$)

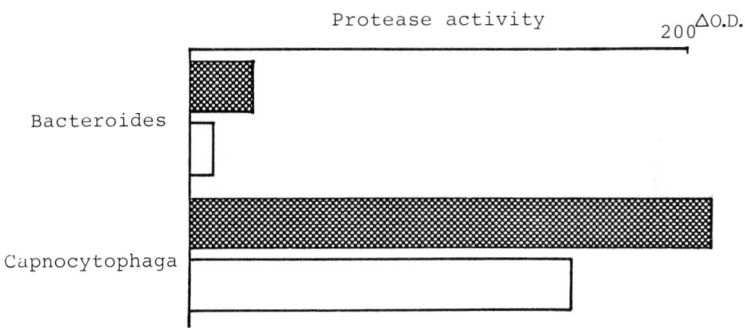

                    Protease activity $200 ^{\Delta O.D.}$

Figure 3.   Protease activity of bacterial cells isolated from
the calculus group(▨) and the non-calculus group(☐)[18] .

species(Fig. 3).   It was concluded that the protease from
Capnocytophaga might play an important role in calculus formation.

A HYPOTHESIS OF CALCIUM PHOSPHATE DEPOSITION ON DENTAL
CALCULUS

   A hypothesis of calcium deposition on dental plaque   was
offered   to explain the relationship between the protease

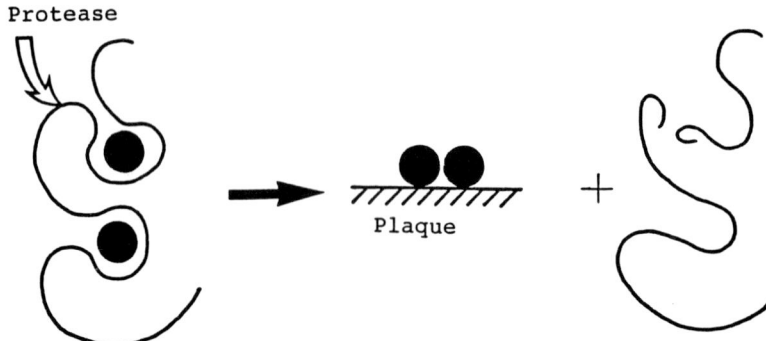

Figure 4. Hypothetical model of a plaque mineralization. Calcium phosphate(●) precipitation can be occurred as a result of degradation of the precipitation inhibitors.

Table 9.EFFECT OF PROTEASE ON CALCIUM-PHOSPHATE PRECIPITATION[22]

Parotid saliva	Purified enzyme	Iodoacetic acid	Precipitation phosphate (mM)
−	−	−	2.55 ± 0.15
+	−	−	0
+	+	−	2.20 ± 0.14
+	+ $	−	0
+	+	+	0.27 ± 0.41

n=5, $ Heat denatured enzyme

activity and calculus formation[7]. Statherin[19]and proline-rich protein[20] have been reported as the inhibitors of calcium phosphate precipitations in human saliva. The degradation of these proteins results in complete loss of the activity[21]. It is very tempting to suppose that these inhibitors are decomposed by salivary protease, and calcium phosphate is easily precipitated onto supragingival calculus.

Bacteria with caseinolytic activity were isolated from the whole saliva of a patient with heavy dental calculus[22]. A protease was purified from a microorganism identified as Bacteroides loescheii (Bacteroides melaninogenicus subsp. melaninogenicus). The purified enzyme was incubated with parotid saliva in the supersaturated calcium phosphate solution. The amount of phosphate precipitated was measured (Table 9). No precipitation was observed in the supersaturated solution with parotid saliva. Addition of the purified enzyme to the solution revealed the phosphate precipitation after the incubation.

Calcium phosphate precipitation is interrupted by the inhibitors of the proteins in saliva, while saliva is supersaturated solution of calcium phosphate. Proteases derived from the microorganisms as Capnocytophaga in the dental plaque degrade the inhibitors, and calcium phosphate precipitation is easily occurred on the plaque. The results suggest that the proteolytic degradation of the inhibitors by oral bacteria is one of the possible mechanism for calcium phosphate precipitation on dental calculus(Fig. 4).

REFERENCES

1. Jensen, A.L. (1959) J. Ammer. Dent. Assoc., 59 , 923-930.
2. Demmers, D.C. and Belting, C.M.(1967) J. Periodontol., 38, 294-301.
3. Schroeder, H.E.(1969) Formation and Inhibition of Dental Calculus. 3.3 Attempts at Dental Calculus Inhibition, p136 -139, Hans Huber Berne, Switzerland.
4. Green, J.C. and Vermillion, J.R.(1960) J. Ammer. Dent. Assoc., 61, 172-179.
5. Massler, M., Schour, I. and Chopra, B.(1950) J. Periodontol., 21,146-164.
6. Watanabe, T., Shinmoto, M., Toda, K., Morishita, M. and Iwamoto, Y. (1980) J. Dent. Res., 59, 138.
7. Watanabe, T., Toda, K., Morishita, M. and Iwamoto, Y. (1982) J. Dent. Res.,61, 1048-1051.
8. Tzamouranis, A., Mathys, J., Ishikawa, I. and Cimasony, G.(1977) Arch. Oral. Biol., 22, 375-378.
9. Söder, P.-Ö. (1972) J. Dent. Res.(Suppl.) 51, 389-393.
10. Cowman, R.A., Perrelal, M.M. and Fitzgerald, R.J.(1976) J. Dent. Res., 55, 391-399.
11. Watanabe, T., Ohata, N., Morishita, M and Iwamoto, Y.(1981) J. Dent. Res., 60,1039-1044.
12. Frostell, G., Nord, C.E. and Söder, P.-Ö.(1973) Odont. Rev. 24, 27-38.
13. Orban, B. and Weinmann, J.P.(1939) J. Ammer. Dent. Assoc., 26, 2008-2017.
14. Watanabe, T., Ohata, N., Morishita, M. and Iwamoto, Y.(1980) Jap. J. Periodontol., 22, 246-251, in Japanese.
15. Tokumoto, K., Ohata, N., Watanabe, T. and Iwamoto, Y.(1982) J. Hiroshima Univ. Dent. Soc., 14, 11-14, in Japaneae.
16. Klinkhamer, J.M.(1968) Periodontics, 6, 253-256.
17. Morita, M. and Watanabe, T.(1986) J. Dent. Res., 65,703-705.
18. Morita, M., Kimura, T. and Watanabe, T.(1988) J. Dent. Res. submitted.
19. Schlesinger,D.H. and Hay, D.I.(1977) J. Biol. Chem., 252, 1689- 1695.
20. Hay, D.I. and Moreno, E.C. (1979) in Proceedings: Saliva and Dental Calculus,(ed. Kleinberg, I., Elickson, S.A. and Mandel, I.D.) pp45-58. Sp.Supp. Microbiology Abstracts, Stoney Book, NY.
21. Hay, D.I. and Grøn, P. (1976) in Proceedings; Microbiology Aspects of Dental Caries,(ed. Stiles, H.M., Loesche, E.J. and O'Brien, T.C.) pp143-150. Sp. Supp. Microbiology Abstracts, Vol. I, Washington, D.C..
22. Morishita, M., Tokumoto, K., Watanabe, T. and Iwamoto, Y. (1986) Archs. Oral Biol., 31, 555-557.

G.Embery

The organic matrix of dental calculus and its interaction with mineral

Department of Basic Dental Science, The Dental School, University of Wales College of Medicine, Heath Park, Cardiff CF4 4XY, UK

ABSTRACT

The mechanisms by which the mineralisation of dental calculus is initiated are not fully understood but it is felt that in keeping with the physiological events operating in other mineralised systems, the organic components must play a vital part. The organic phase of dental calculus contains a wide variety of components of salivary, cellular and bacterial origin. The protein residues are characterised by high contents of glycine, alanine, glutamate and aspartate and low contents of methionine and cystine. Neither cystine nor hydroxyproline are present, suggesting the absence of keratin or collagen. The lipid fractions comprise free fatty acids, phospholipids, cholesterol esters and triglycerides. Sulphated glycopeptides of salivary origin are present, together with acidic glycosaminoglycans such as hyaluronic acid, dermatan sulphate and chondroitin 4-sulphate. Differences between the composition of subgingival and supragingival calculus are evident with certain organic components of the latter being derived via the gingival sulcus exudate from associated destruction of periodontal tissue. Studies have been made between the interaction of the organic and inorganic phases using infrared spectroscopy, adsorption curves and equilibrium dialysis. Using enzymic digestion of salivary glycoconjugate-hydroxyapatite interactions the minimum apatite-binding regions were shown to be N-acetyl-hexosamines which classically bear the sulphate ester residues. It is evident that polyanionic molecules have an important role to play in calculus formation presumably, as nucleating sites for Ca and PO_4 ion deposition from metastable systems.

Recent Advances in the Study of
Dental Calculus

INTRODUCTION
    The mechanisms which initiate the ectopic mineralisation of
tooth surface deposits to form dental calculus are still
incompletely understood. In contrast to the molecular events
which occur in the homeostasis of bone and dentine, dental
calculus represents a surface phenomenon and is therefore
outside the normal cellular, neural and hormonal controlling
influences. As such there is no remodelling and calculus will
form and increase in bulk via accretion due to surface activity.
Limits on its size will be imposed by shear forces in the mouth
particularly by fragmentation due to masticatory forces.
    Along with the calcification systems which occur in
physiological systems, the organic components present are
considered to play an important role in the mineralisation
process. Approximately 20% of dental calculus is organic
matrix, the major part being derived from saliva, gingival
sulcus fluid and oral bacteria with a contribution from cell
debris such as polymorphonuclear leucocytes.

EARLY STUDIES ON ORGANIC COMPONENTS
    It has always been considered that dental calculus forms
through the mineralisation of dental plaque which forms on the
teeth in the absence of adequate dental hygiene. Nevertheless
it has been known for some years that dental calculus can form
on the teeth of germ-free animals, a finding which pointed to a
prime role for salivary organic components in the mineralisation
process, particularly salivary protein-carbohydrate complexes
such as anionic glycoproteins.
    In 1957 Mandel and Levy (1) suggested a role for
protein-polysaccharide complexes in the formation of calculus.
These workers demonstrated the presence of carbohydrate-protein
molecules in supragingival calculus using chemical and
histochemical techniques and in subgingival calculus using
histochemical procedures alone. Further studies by this group
and by Little et al (2) yielded valuable information on the
carbohydrate and amino acid constituents of acid hydrolysates of
the supragingival calculus from the molar and lingual areas of a
large number of individual samples.

AMINO ACID CONTENT
    The most thorough analysis of the protein constituents of
dental calculus in recent years was carried out by Osuoji and
Rowles (3) when they investigated the amino acid content of
demineralised acid hydrolysates of mixed supra and subgingival
calculus and subgingival calculus separately. The results of
the amino acid content are given in Table 1.
    Seventeen amino acids were detected with glutamic, aspartic,
glycine, alanine, valine and leucine forming the largest
proportion of the total residues isolated. The sulphur
containing amino acids, methionine and cysteine were present in
only trace amounts. Neither cystine or hydroxyproline were
detected suggesting the absence of keratin or collagen.
    This information agrees closely with the earlier work of
Little et al (2) who additionally noted differences between the
lower lingual and upper molar calculus samples where the amino
acid concentration of the former was found to be lower and of a

Table 1. AMINO   ACID   COMPOSITION   OF   DENTAL   CALCULUS   AND
                              SUBGINGIVAL CALCULUS
                 (values expressed as residues per 1000 total residues
                 corrected to whole numbers above 10.0)

Amino acid	Dental calculus	Subgingival calculus
Lys	39	34
His	4	3.7
Arg	32	23
Asp	115	150
Thr	52	72
Ser	55	71
Glu	111	128
Pro	48	43
Gly	131	102
Ala	105	121
Cys/2	4.9	3.2
Val	68	79
Met	19	14
Ileu	50	42
Leu	83	71
Tyr	24	16
Phe	58	29

different composition. The nitrogen content of the upper molar
sample was also consistently higher than that of the lower
anterior calculus.

    Such information yields important clues to the differences
and origin of calculus components at different sites of the
mouth and clearly reflects the proximity of the various salivary
duct openings. It tells us very little about the nature of the
constituent protein derivatives which will be dealt with in a
later section.

LIPID CONTENT
    Few detailed accounts of the lipid fraction have yet been
carried out although it is known that the lipid content of
saliva and certain glycoprotein fractions is significantly high.
The lipid content of dried calculus was studied by Osuoji and
Rowles (3) using chloroform-methanol extraction. A combination
of thin-layer chromatography and gas-liquid chromatography
showed the presence of a variety of lipid components and fatty
acids. The lipid content was 15.3% of the dry weight of the
decalcified calculus and included phospholipids, cholesterol
esters, diglycerides, triglycerides and free fatty acids.

    The free fatty acids represented the largest component of the
lipid fraction and the predominant acids detected were palmitic,
stearic and oleic with smaller amounts of the unsaturated fatty
acids, linoleic and linolenic. The nature and composition of
the parent lipids and their origin is not known although it is
considered likely that bacterially derived lipopolysaccharides
and phospholipids are candidates.

    Slomiany and his coworkers (4) have shown that the organic

Table 2.   CARBOHYDRATE ANALYSIS OF THE ORGANIC MATRIX OF DENTAL
            CALCULUS (% dry wt.)

Sugar	Dental calculus	EDTA-soluble matrix
Hexuronic Acid	2.98	0.92
Hexosamine	5.48	1.15
Glucose	5.65	0.49
Galactose	4.91	0.92
Mannose	3.14	0.01
Rhamnose	7.62	0
Xylose	0	0.32
Arabinose	3.48	0

matrix of supragingival calculus contains up to 15% of lipid. Neutral lipids represent 62% of the content with glycolipids accounting for a further 28%. Phospholipids account for 10% of the total lipid and include phosphatidylethanolamine and diphosphatidylglycerol.

Takazoe et al (5) showed the a lipid extract from the calcifiable micro-organism Bacterionema matruchotii causes in vitro nucleation of hydroxyapatite. In a further study a proteolipid isolated from dental calculus was also shown to cause nucleation of a metastable calcium phosphate solution.

CARBOHYDRATE CONTENT

Studies by Little et al (2) on acid hydrolysates of decalcified calculus indicated the presence of a variety of constituent sugars including glucose, galactose, mannose, glucuronic acid, galacturonic acid, glucosamine, galactosamine, rhamnose and sometimes arabinose. It was noteworthy that in this study fucose, xylose, deoxyribose, raffinose and sialic acid were not found. No differences in carbohydrate content either chemically or chromatographically were found between calculus sites or different mouths.

Similar findings were observed by Osuoji and Rowles (3) and their results are shown in Table 2.

The constituent sugars yield some information on the source of origin of the calculus components. Hexuronic acid is notably a component of glycosaminoglycans present in connective tissue. It is not found in saliva or in the cell walls of oral micro-organisms (8). It is however a basic component of capsular streptococci and slimes (9) where it is present as hyaluronic acid and other unidentified uronides.

The absence of deoxyribose and ribose preclude the presence of nucleic acids in calculus and indicates that the oral micro-organisms undergo extensive degradation leaving only the cell wall for calculus formation.

Rhamnose was the principal sugar detected and its presence indicates a bacterial origin for at least part of this sugar. The absence of sialic acid and fucose suggests that only part of the salivary glycoprotein residues remain for calculus formation (3).

MACROMOLECULES IN DENTAL CALCULUS

Very few studies have dealt with the isolation of intact macromolecules from calculus. Conceivably most salivary components particularly those with ionisable groupings will be able to interact with the amphoteric surface of calcium phosphate during its maturation stage although may not necessarily be involved during the initial nucleation or seeding stage. A variety of compounds have been detected in plaque and pellicle and some of these constituents either in their native or modified form could be present in calculus.

(a) Salivary glycoproteins. Histochemical studies and carbohydrate and amino acid profiles have all implicated protein-polysaccharides or mucoid substances in the structure of calculus. Embery (7) reported the presence of a sulphated glycopeptide in supragingival calculus which contained a high amount of ester sulphate, hexose, equal amounts of galactosamine and glucosamine and a small quantity of peptide. Hexuronic acid and sialic acid were not detected. The compound had clearly undergone a major transformation from its salivary native form either prior to or during calculus maturation. Infrared analysis indicated that the sulphate group was covalently bound at the carbon-6 position of hexosamine.

(b) Glycosaminoglycans. The presence of hexuronic acid in calculus samples has been known for some years. Although uronides are present in certain bacteria their detection as glycosaminoglycans in calculus has been considered unusual. Earlier studies (10, 11) have shown that hyaluronic acid, chondroitin sulphate and possibly heparan sulphate are present in mixed supragingival calculus. Studies by Embery and Whitehead (13) and Embery and Nordbo (13) have indicated that the non-sulphated glycosaminoglycan, hyaluronic acid, is present in supragingival calculus whereas the sulphated glycosaminoglycans chondroitin 4-sulphate and dermatan sulphate are additional components in subgingival calculus.

The sulphated glycosaminoglycans arise out of periodontal tissue breakdown via the gingival crevicular fluid (14). The glycosaminoglycans generally are associated with tissue structure and contribute to mineralisation in dentine and bone due to their highly charged nature arising out of the carboxyl and ester sulphate groups. They therefore make ideal candidates for ectopic mineralisation where they could feature as nucleating sites for calcium phosphate deposition.

(c) Phosphoproteins. During recent years it has emerged that phosphoproteins represent important constituents of human saliva. Proline-rich phosphoproteins represent 28% of the proteins in human parotid and submandibular saliva. Such compounds are detected in pellicle and are known to undergo metabolic breakdown (15) a finding supported by loss of phosphoprotein or certainly the phosphate anion determinant with formation time (16). An immunochemical study of host proteins in human supragingival plaque indicated that cysteine-containing phosphoproteins were present in plaque but that the proline-rich phosphoproteins were virtually absent.

It is also of interest that the phosphoprotein, statherin, is present in saliva and able to control the precipitation of metastable calcium phosphate solutions in vitro (17).

(d) Proteolipid and phospholipid. A phospholipid fraction was obtained from crude calculus and fractionated by exchange chromatography on hydroxy-propylated dextran gel. One fraction identified as a proteolipid was able to induce hydroxyapatite formation from a metastable calcium phosphate solution. A similar fraction has been isolated from Bacterionema matruchotii and proved to be a calcification nucleator of the micro-organism and may have this role in bacterial-induced calculus formation.

(e) Collagen and keratin. Neither of these molecules nor their degradation products have been identified in calculus, a finding confirmed by the absence of hydroxyproline and cysteine from the amino acid profile.

INTERACTION BETWEEN THE ORGANIC AND INORGANIC PHASES OF DENTAL CALCULUS

The general theme throughout this review is that organic components are important in the initial nucleation phase of dental calculus and play a major role in the subsequent maturation process. There are few reports on the interactive mechanisms between the organic components and the calcium phosphate phase.

The problem has been approached by ourselves and other workers using hydroxyapatite as the mineral support medium. This is the major crystalline form in mature calculus ($Ca_{10}(PO_4)_6 OH_2$) with smaller quantities of octacalcium phosphate ($Ca_8(HPO_4)_4$), whitlockite, a magnesium calcium phosphate ($Ca_3(PO_4)_2$) and brushite ($CaHPO_4 2H_2O$). Brushite is the major form in calculus deposits less than three months old.

(a) Phosphoproteins. The interaction between salivary phosphoproteins and hydroxyapatite has been studied by Bennick et al (18). Phosphoproteins rich in proline represent an important aspect of pellicle and possibly calculus formation but do not retain their parent structure in situ. Attachment of the phosphoproteins obtained from human saliva to hydroxyapatite followed by selective enzyme digestion of the complex indicates that the binding site is located in the proline-poor N-terminal part of the protein possibly located between amino acid residues 3 and 25 which is also the site of calcium attachment. Phosphoserine is also necessary for attachment. It is also known that these proteins are able to prevent the transformation of metastable calcium phosphate into hydroxyapatite (17).

(b) Glycosaminoglycans. The interaction between glycosaminoglycans and hydroxyapatite has been studied by Embery, Rolla and Stanbury (19). The order of binding of selected glycosaminoglycans was heparin, heparan sulphate, chondroitin sulphates and hyaluronic acid indicating that the degree of negative charge along the carbohydrate chain which arises from carboxyl and sulphate groups was the major binding determinant. The interaction presumably takes place

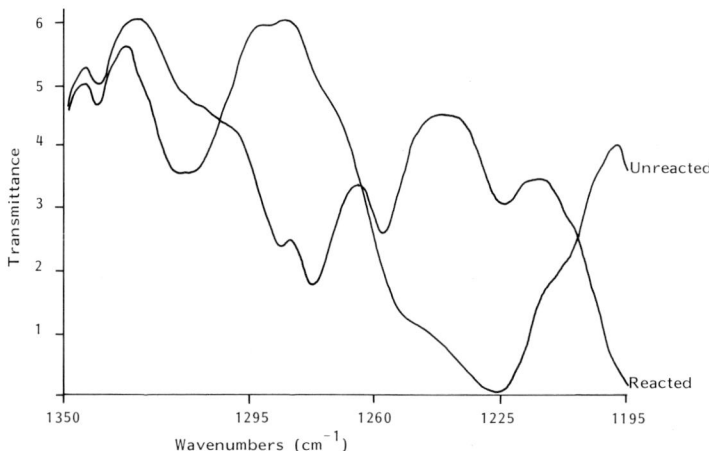

Figure 1. Interactive infrared spectra between salivary sulphated glycoprotein and hydroxyapatite.

electrostatically through calcium sites on the crystal lattice, a postulate enhanced by the observation that calcium pretreatment increases binding whereas fluoride leads to a decrease.

Infrared spectroscopy has revealed that the sulphate groups are directly involved in the interaction of chondroitin 4-sulphate since the S=0 vibrations at 1250 and 850cm^{-1} are lost. It is also notable that the binding level of chondroitin 4-sulphate is increased when hydroxyapatite is pretreated with saliva ((20).

(c) <u>Salivary glycoproteins</u>. Salivary glycoproteins bind strongly to hydroxyapatite and are implicated in calculus calcification. The mechanism of interaction of a sulphated glycoprotein with hydroxyapatite has been studied recently by ourselves using FTIR infrared spectroscopy and selective enzyme digestion. The sulphated glycoprotein isolated from human saliva contained 30% protein, 52% carbohydrate and 8% ester sulphate with the sulphate present as N-acetylgalactosamine-6-0-sulphate.

There were spectral changes on reaction with hydroxyapatite at 1230, 775 and 828cm^{-1} due to the involvement of sulphate groups. Evidence of amide change in the regions of 1670 and 1550 cm^{-1} is also apparent due to attachment of the polypeptide region. It is proposed that the exposed parts of the protein core are involved in the attachment process with the main interactive residues being the sulphate groups present on the carbohydrate apo-protein moiety (2).

The spectra for reacted (test) and unreacted (control) samples are shown in Fig. 1 showing the reduced absorption at 1230 cm^{-1} in the reacted mixture.

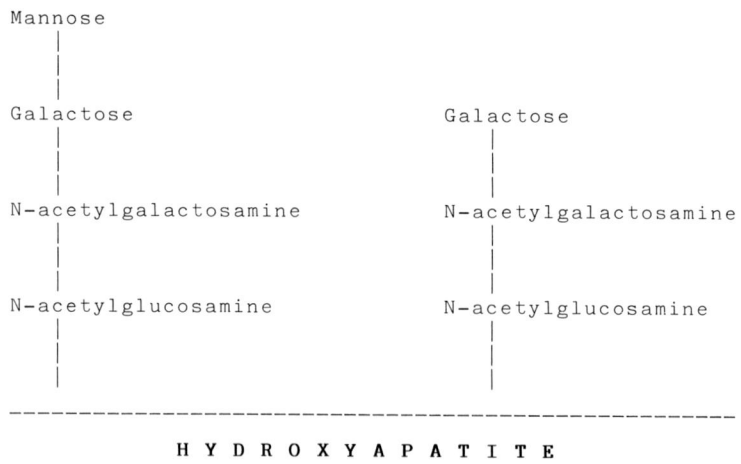

```
Mannose
   |
   |
   |
Galactose                          Galactose
   |                                   |
   |                                   |
   |                                   |
N-acetylgalactosamine              N-acetylgalactosamine
   |                                   |
   |                                   |
   |                                   |
N-acetylglucosamine                N-acetylglucosamine
   |                                   |
   |                                   |
   |                                   |
```

H Y D R O X Y A P A T I T E

Figure   2.   Hydroxyapatite-binding   regions   of   salivary
glycoprotein.

    A further approach used by ourselves has been an attempt to
characterise the hydroxyapatite-binding regions of the sulphated
glycoprotein.   This was achieved by enzymic digestion with
proteolytic   enzymes   and   fucosidase   of   sulphated
glycoprotein-hydroxyapatite complexes.   The minimum molecular
size regions remaining attached to the hydroxyapatite following
extensive digestion are shown in Fig. 2 and are less than 1000
daltons.
    Sialic acid is also involved in the attachment process as are
certain regions of the polypeptide chain as evidenced by the
presence of amino acid residues in the fractions associated with
the minimum binding regions.   It is our contention that the
hexosamine sulphate residues of the glycoprotein side-chains in
association with certain peptide and terminal sialic acid
residues represent the hydroxyapatite-binding regions (22).

THE ROLE OF THE ORGANIC PHASE IN MINERALISATION
    Many reviews have dealt with calculus formation and its
prevention, the most recent being an article by Mandel (23).   A
number of theories have been put forward including the 'booster'
mechanism   where   the   local   pH,   calcium   and   phosphate
concentrations are high leading to calcium phosphate deposition,
the   epitactic   theory   where   metastable   calcium   phosphate
solutions seed and nucleate from a suitable template and the
inhibition theory where a potential calcification inhibitor is
removed eg. pyrophosphatase removal of inhibitory pyrophosphate.
    It is more than likely that all of the factors occur at one
and the same time.   Nevertheless the role of the organic phase
cannot be overlooked at all stages of calculus growth.   It is
probable that the initial seeding events may be different from
the subsequent maturation events.

Current views hold that calculus can form in germ-free animals although the contribution of bacteria cannot be ignored when they are present.

Ennever et al (24) contend that the initiator of calculus matrix calcification is a proteolipid. A similar molecule promotes nucleation by the micro-organism Bacterionema matruchotti.

The presence of glycosaminoglycan-rich proteoglycan metabolites in calculus are also potential nucleators due to their high number of sulphate and carboxyl groups which can attract calcium. Another important feature of these molecules which also closely resemble the anionic salivary glycoproteins is that in addition to concentrating calcium ions they may attract water molecules and thereby locally increase the metastability of calcium phosphate solution leading to deposition around the organic macromolecules which then feature as templates.

Additionally such molecules can inhibit or certainly regulate calcium phosphate formation. Chen and Boskey (25) have shown that coating hydroxyapatite seed crystals with proteoglycan, chondroitin sulphate and dextran sulphate decreased the amount of hydroxyapatite precipitated as a function of time whereas the desulphated analogues showed less inhibition. On the other hand the removal of sulphate may induce conformational changes indicating that charge may not be the only determinant.

The shielding of the negative sites on the hydroxyapatite surface may be one controlling determinant. In vivo the metabolism of such molecules must not be overlooked since protease and glycosidase enzymes are present in high amounts. Protease enzyme has been identified in relation to dental plaque and supragingival calculus formation and collagenase is known to interact with hydroxyapatite (25). The presence of enzymes on the surface of calculus may also be important in controlling its formation particulary since enzymes increase their turnover activity in the immobilised form.

PREVENTIVE ASPECTS

Attempts to reduce the level of calculus formation have been the subject of much research for the past 30 years (23). Various inhibitory agents including enzymes, vitamin C, urea and certain antibiotics have been examined.

In the last decade however emphasis has moved towards inhibitors that control crystal growth. These include diphosphonates and more recently ethylene diamine tetramethylene phosphonic acid and sulfoacrylic acid (23). There is much renewed interest in the use of zinc salts as anticalculus agents. Information on the use of zinc salts will be the subject of other reviews in this book. However, although it is known that zinc binds to hydroxyapatite and inhibits crystal growth of the mineral little attention has been given to the fact that zinc and other inhibitory agents may block sites on the organic matrix in vivo thus reducing or altering the template form and masking potential nucleation sites.

It is contended that only when a thorough understanding of the organic components of calculus and their interaction with the mineral phase is gained will an absolute preventive strategy be formulated.

REFERENCES
1. Mandel, I.D. and Levy, B.M. (1957) Oral Surg. 10, 874-884.
2. Little, M.F., Bowman, L., Casciani, C.A. and Rowley, J. (1966) Archs. Oral Biol. 11, 385-386.
3. Osuoji, C.I. and Rowles, S.L. (1974). Calcif. Tiss. Res. 16, 193-200.
4. Slomiany, B.L., Murty, V.L.N., Aono, M., Sarosiek, J., Slomiany, A. and Mandel, I.D. (1983) J. Dent. Res. 62, 862-865.
5. Takazoe, I., Vogel, J.J. and Ennever, J. (1970) J. Dent. Res. 49, 395-398.
6. Mandel, I.D., Thompson, R. and Ellison, S.A. (1964) Archs. Oral Biol. 9, 601-609.
7. Embery, G. (1977) Calc. Tiss. Res. 23, 13-17.
8. Luria, S.E. (1960) The Bacteria. A treatise on structure and function. Vol. I. Structure. (Eds. by Gunsalus, I.C. and Stainer, R.Y.) Chapt. I, pp. 1-34 Academic, New York
9. Salton, M.R. (1960) In "Structure" (Eds. Gunsalus, I.C. and Stainer, R.Y.) Vol. 1., Chapt. 3, Academic Press. 97-151.
10. Paunio, K.U. and Paunio, I.K. (1970). Scand. J. Dent. Res. 78, 447-451.
11. Osuoji, C.I. and Rowles, S.L. (1972) Archs. Oral Biol. 17, 211-214.
12. Embery, G. and Whitehead, E. (1976) Calc. Tiss. Res. 22, 227-229.
13. Embery, G. and Nordbo, H. (1979) Scand. J. Dent. Res. 87, 325-327.
14. Embery, G., Oliver, W.M., Stanbury, J.B. and Purvis, J.A. (1982) Archs. Oral Biol. 27, 177-179.
15. Bennick, A., Cannon, M. and Madapallimattam, G. (1981) Caries Res. 15, 9-20.
16. Embery, G., Heaney, T.G. and Stanbury, J.B. (1986). Archs. Oral Biol. 31, 623-625.
17. Hay, D.I. and Gron, P. (1976) Archs. Oral Biol. 21, 201-205.
18. Bennick, A., Cannon, M. and Madapallimattam, G. (1979) Biochem. J. 183, 115-126.
19. Embery, G., Rolla, G. and Stanbury, J.B. (1979) Scand. J. Dent. Res. 87, 318-324.
20. Embery, G. and Rolla, G. (1980) Acta Odontol. Scand. 38, 105-108.
21. Embery, G. and Green, D.R.J. (1986) J. Dent. Res. 65, Abstr. 734.
22. Embery, G. and Green, D.R.J. (1986) J. Dent. Res. 65, Abstr. 733.
23. Mandel, I.D. (1988) In "Compend. Contin. Educ. Dent. Suppl." No. 8, 235-241.
24. Ennever, J., Vogel, J.J., Riggan, L.J. and Paoloski, S.B. (1977) J. Dent. Res. 56, 140-142.
25. Chen, C.C. and Boskey, A.L. (1985) Calcif. Tiss. Internat. 37, 395-400.
26. Knuuttila, M.L.E. and Paunio, K.U. (1978) Calcif. Tiss. Res. 25, 127-131.

SESSION III.
Composition and
Formation of Calculus

H.J.Busscher
H.M.W.Uyen
W.L.Jongebloed[1]
L.J.van Dijk

Adhesional aspects of dental calculus formation

Laboratory for Materia Technica, University of
Groningen, Antonius Deusinglaan 1, 9713 AV
Groningen, The Netherlands and [1]Department of
Histology and Cell Biology, University of Groningen,
Oostersingel 69/2, 9713 EZ Groningen,
The Netherlands

ABSTRACT

The adhesional bond strength of calculus to enamel and dentine surfaces determines the ease at which calculus can be removed by daily tooth brushing or professional dental treatment. In this study we examined the adhesion of calculus to various smooth and rough substrata with different surface free energies using a beagle dog model. Whereas the rate of calculus formation was neither greatly influenced by the substratum surface free energy nor by its roughness, removal by brushing and ultrasonic instrumentation was clearly easier from smooth and low surface free energy substrata. The expectation seems justified therefore that application of surface free energy reducing agents to the enamel surface may lead to an easier removal of calculus in vivo.

Recent Advances in the Study of
Dental Calculus

INTRODUCTION

Formation of dental calculus can occur both supragingivally as well as subgingivally and is a great nuisance for patients, especially since calculus can form despite a good oral hygiene. Chronic irritation of periodontal tissue by subgingival calculus can cause destructive periodontal diseases. Often toothbrushing alone is not sufficient to prevent calculus formation and regular scaling by a professional dentist is required.

Scaling, although harmless at first glance, can be detrimental to the enamel and root surface if applied over a long time (1-4). Fig. 1A shows an electronmicrograph of an enamel surface after ultrasonic instrumentation with distinct cavitations due to treatment. It is sometimes estimated that removal of calculus by scaling removes approximately 5-10 µm of sound enamel or dentine. Averaged over 30 years, a reasonable assumption for the duration of a treatment with two yearly visits to a dentist, this indicates that together with calculus 0.6 mm of enamel or dentin is removed too. Fig. 1B shows the clinical appearance of teeth from a patient after many years of regular scaling. Clearly sound tissue has disappeared together with calculus.

 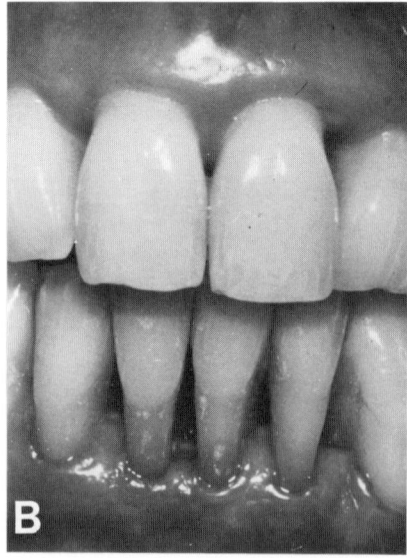

Figure 1.A - SEM-picture of an enamel surface after ultrasonic scaling showing the destructive side effects of instrumented treatment. The bar denotes 100 µm.

Figure 1.B - Clinical appearance of teeth after prolonged instrumented removal of calculus.

As improved oral hygiene does not always seem to be effective in reducing the occurence of dental calculus, preventive measures include the addition of e.g. pyrophosphate (5, 6), zinc citrate (7), hexametaphosphate (8) or disodiumetidronate (9) to toothpastes. Hitherto however the success of these additives has not been ubiquitously accepted.

Since it seems impossible at present to prevent calculus formation, we have recently focussed on the interfacial properties between substratum and calculus in order to investigate whether removal would be easier from specific materials (10). In an in vivo study in beagle dogs, it appeared that although the rate of calculus formation was not influenced by substratum properties, removal, as studied in a brushing machine was much easier from low surface free energy substrata than from high surface free energy substrata.

It is the aim of this paper to determine the combined influence of substratum surface free energy and roughness on the in vivo rate of calculus formation and removal. For this purpose, removal will be studied in a brushing machine, representative for the removal that can be daily accomplished by people and by means of ultrasonic instrumentation as carried out by a professional dentist.

MATERIALS AND METHODS

Set-up of the study
In four beagle dogs, 4-5 years old, fenestrated gold crowns, suitable to contain two facings each were prepared up to the gingival margin in the upper fourth pre-molars and cemented with a non-fluoridated cementum (Nogenol® COE) as described in detail previously (11).

Polytetrafluorethylene (PTFE, γ_s = 20 mJ.m-2), polymethyl-methacrylate (PMMA, γ_s = 56 mJ.m-2), glass (γ_s = 120 mJ.m-2) and bovine enamel (γ_s = 85 mJ.m-2) samples were cut into facings of 5 x 5.5 x 1 mm. Smooth surfaces (stylus surface roughness R_a ≈ 0.1 µm) were obtained by polishing, whereas samples were roughened by sand blasting. The different substrata were placed in the gold crowns for various time intervals ranging from 1 to 28 days. During the experimental period, the dogs were fed with soft food containing, however no special calcium rich components. Occasionally SEM-pictures were made of the specimens.

Evaluation of the removal of calcium-rich deposits
After removal of the substrata from the oral cavity after 1, 3, 7 , 14 or 28 days the substrata were rinsed in physiological saline and extensively washed in distilled water to remove loosely bound material and stained for 2 minutes with Alizarin Red S, a specific stain for calcium, not giving any discolouration of enamel, pellicle or non-mineralized plaque (12). After staining, the initial percentage of the substratum surface covered with calculus was determined from enlarged photonegatives using a Graphics Tablet (Apple Computer Inc.).

Subsequently, the substrata were put in a brushing machine (13), equiped with a medium stiff brush (Oral B35). Brushing was carried out in distilled water, exerting a force of 151 gram. The number of strokes required to reduce the surface coverage by calcium rich, stainable deposits with 63% was taken as a measure for the strength of adhesion between calculus and substratum.

As an alternative measure for the strength of adhesion, the width of calculus removed by manually moving a TFI-10 insert of a Cavitron® 2002TM (Dentsply® International Inc.) ultrasonic device operated at medium power setting over the adhering calculus layer was taken.

Evaluation of the adhesion of calcium-rich deposits

For gravimetric evaluation of calcium rich deposits, the weight of the samples was determined prior to insertion in the oral cavity and after removal with a precision of 0.2 mg.cm-2. Previously we have shown that the weight of a sample becomes contant after 24 hours drying in air (10). All weights of the calcium rich deposits on the various materials have therefore been determined in fourfold after 24 hours drying of the samples.

RESULTS

Table 1 lists the formation rate of calium rich deposits on smooth and rough substrata with varying surface free energies. As can be seen, there is no significant influence of substratum surface free energy nor of surface roughness on the formation rate, although slightly less deposit seems to be formed on the low surface free energy PTFE surface, in accordance with previous data (10).

It should be noted, that under the present experimental conditions the roughness R_A of the calculus surface itself becomes increasingly high in time, viz. 0.5, 1.2, 2.1, 3.4 and 5.3 µm after 1, 3, 7, 14 and 28 days respectively. However, brushing without toothpaste rapidly decreased the calculus surface roughness to approximately 1.8 µm. If the calculus surface was brushed with toothpaste, its roughness was reduced to about 0.5 µm, normal for an in vivo tooth surface too.

Fig. 2 shows SEM-images of calculus formed on a smooth (2A) and rough (2B) high surface free energy glass surface, a smooth enamel surface (2C) and a rough low surface free energy PTFE surface (2D). Two major differences appear between calculus formed on low and high surface free energy materials, viz.
1. a less compact calculus structure on the low surface free energy material
2. a much closer contact between high surface free energy materials and calculus.

Additionally, deposits remain on the rough and high surface free energy substrata after freeze-drying and fracturing the specimens for SEM preparation, indicative for a high adhesional bond strength.

The efficiency of removal of calculus by brushing is illustrated in Table II averaged for deposits formed during 1 and 2 weeks.

Whereas the age of the substratum-calculus interface does

Figure 2. Electron micrographs of calculus formed on various smooth and rough materials. A. smooth glass. B. rough glass. C. smooth enamel. D. rough PTFE-teflon. The bar denotes 10 µm.

not influence the ease of calculus removal by brushing, removal is clearly faster accomplished on smooth, low surface free energy materials.

The efficiency of ultrasonic instrumented removal is qualitatively presented in Fig. 3, showing the widths of one single tracing with a Cavitron insert over calculus formed on various substrata.

Table I. Dry weights of calcium rich deposits formed on various
smooth and rough substrata with different surface free
energies exposed to the oral cavity of beagle dogs.
Formation rates ($\pm$ 20% S.D. over four dogs) were
obtained by a linear least square fit over a period of
28 days.

substratum	γ_S (mJ.m-2)	R_A (µm)	formation rate (mg.cm-2.day-1)	linear correlation coefficient
PTFE	20	0.07	0.11	0.95
		0.75	0.18	0.96
PMMA	56	0.07	0.20	0.98
		0.65	0.14	0.89
Glass	120	0.07	0.19	0.996
		1.35	0.18	0.99
bovine enamel	85	0.06	0.22	0.98
		1.32	0.23	0.99

Table II. Number of strokes required to remove 63% of
stainable, calcium rich deposits from smooth and
rough substrata with different surface free energies.

substratum	smooth	rough
PTFE	20	40.000
PMMA	20	>40.000
Glass	>20.000	>40.000
bovine enamel	>20.000	>40.000

    Although at present, the technique for quantification of
ultrasonic calculus removal is not ideal and can be improved by
using constant force and speed control devices, we have
attempted to quantitatively analyse these results by measuring
the widths of the Cavitron tracing on the rough substrata only
(see Table III). These results confirm the previous findings
that removal of calculus is easier from low surface free energy
substrata.

DISCUSSION

In this study we examined adhesional aspects of dental calculus
formation on various substrata in vivo using a beagle dog
model.

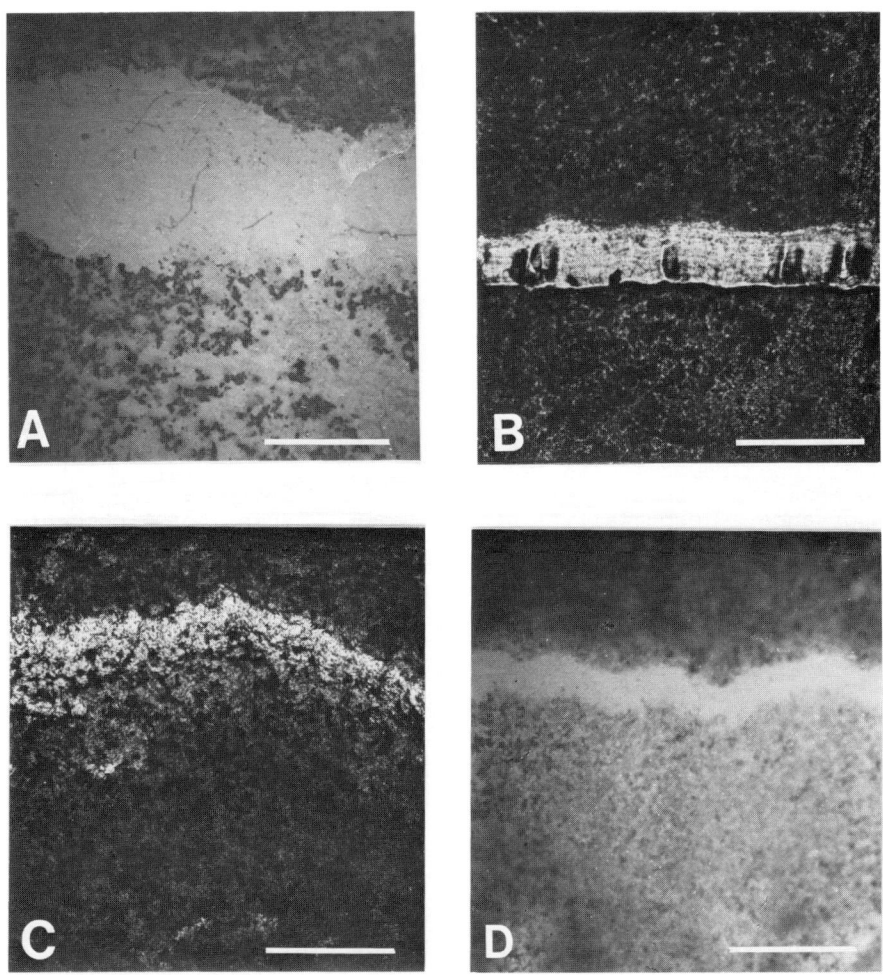

Figure 3. Stainable, calcium rich deposits formed on roughened PTFE (A), PMMA (B), Glass (C) and bovine enamel (D) after one tracing with the insert of a Cavitron. The bar represents 1 mm.

It is important to emphasize at this point that two major differences exist between the beagle dog model employed and the human oral cavity regarding calculus formation. Firstly, calculus formed in beagle dogs is known to possess a high amount of calcium carbonates, whereas human calculus consistently contains calcium phosphates (14). Secondly, as becomes obvious from this study, the calculus surface in the dog model is rougher than generally observed in humans due to the lack of

Table III. Removal widths (± S.D. over four dogs) of Cavitron tracings over stainable, calcium rich deposits formed on various rough substrata.

substratum	removal width [mm]
PTFE	1.27 ± 0.43
PMMA	0.53 ± 0.26
Glass	0.50 ± 0.36
bovine enamel	0.65 ± 0.52

regular brushing. As a rougher surface is more prone to bacterial adhesion (15), this aspect may lead to increased formation rates in dogs under the present experimental conditions compared to man. From an interfacial point of view however, these differences are probably of minor importance, especially since most types of ionogenic crystalline materials all have a high surface free energy (16).

The average amount of deposit formed per cm^2 increased linearly with time up to at least 28 days for both smooth as well as rough surfaces. Only on smooth PTFE a minor deviation from linearity was observed already after 12 to 14 days due to a weaker binding strength of the deposits yielding occasional removal by oral shear forces. In addition, only on PTFE an increased formation rate was observed after sand blasting.

This study describes a first set-up to quantify the removal of calculus by brushing or ultrasonic scaling. Most currently used removal parameters as the RCI "Remaining Calculus Index" or the SRI "Scanning Roughness Index " (2) cannot be considered as real quantitative parameters. The firmer adhesion of calculus to high surface free energy materials as evidenced by brushing and ultrasonic instrumentation can be explained on basis of the high surface free energies of both interacting surfaces, which is a thermodynamically favourable situation for adhesion (17). From a clinical point of view it is important to realize that calculus removal can be facilitated by polishing and surface free energy reduction, since easier removal leads to a smaller loss of sound tissue during instrumented treatment. The quantitative parameters proposed seem more suitable to study removal than the crude clinical indices currently employed based on a 0, 1, 2, 3 score.

A major requirement for prevention of calculus formation is a smooth tooth surface. Once this is achieved, calculus formation may admittedly not be completely inhibited but both daily tooth brushing as well as professional cleaning will most likely yield an easier and better removal as can be concluded from this study. Furthermore, as a second preventive measure, the tooth surface free energy may be reduced by adsorption of aminefluorides (18) or perfluoraklyl surfactants (19) to facilitate an easier removal.

ACKNOWLEDGEMENTS

The authors are greatly indebted to Mrs. Marjon Schakenraad-Dolfing for preparing this manuscript and to Mr. D.M. Meyer from CDL, R.U. Groningen for his assistance with handling the dogs.

REFERENCES

1. Gellin, R.G., Miller, M.C., Javed, T., Engler, W.O. and Mishkin, D.K. (1986) J. Periodontol. 57, 672-680.
2. Lie, T. and Leknest, K.N. (1985) J. Periodontol. 56, 522-531.
3. O'Leary, T.J. (1986) J. Periodontol. 57, 69-75.
4. Breininger, D.R., O'Leary, T.J. and Blumenshine, R.V.H. (1987) J. Periodontol. 58, 9-18.
5. Lobene, R.R. (1986) Clin. Prev. Dent. 8, 5-7.
6. Schiff, T.G. (1987) Clin. Prev. Dent. 9, 13-16.
7. Lobene, R.R. (1987) Clin. Prev. Dent. 9, 3-8.
8. Draus, F.J., Lesniewski, M. and Miklos, F.L. (1970) Archs. Oral Biol. 15, 893-896.
9. Sturzenberg, O.P., Swancar, J.R. and Reiter, G. (1971) J. Periodontol. 42, 416-419.
10. Uyen, H.M., Dijk, L.J. van and Busscher, H.J. (1989) J. Clin. Periodontol. (in press).
11. Dijk, L.J. van, Herkströter, F., Busscher, H.J., Weerkamp, A.H., Jansen, H. and Arends, J. (1987) J. Clin. Periodontol. 14, 300-304.
12. Schroeder, H.E. (1969) in Formation and Inhibition of Dental Calculus, (ed. Huber, H.) pp. 65-622. Hans Huber Publisher Berne, Stuttgart, Vienna.
13. Slop, D., Rooij, J.F. and Arends, J. (1983) Car. Res. 17, 242-248.
14. Legeros, L.Z. and Shannon, I.L. (1979) J. Dent. Res. 58, 2371-2377.
15. Quirynen, M., Marechal, M., Busscher, H.J., Weerkamp, A.H., Darius, P.C. and Steenberghe, D. van. J. Clin. Periodontol. (submitted).
16. Busscher, H.J., Jong, H.P. de and Arends, J. (1987) Mat. Chem. Phys. 17, 553-558.
17. Busscher, H.J., Weerkamp, A.H., Mei, H.C. van der, Pelt, A.W.J. van, Jong, H.P. de and Arends, J. (1984) Appl. Envir. Microbiol. 48, 980-983.
18. Busscher, H.J., Uyen, H.M., Jong, H.P. de and Arends, J. (1988) J. Dent. 16, 166-171.
19. Gaffar, A., Esposito, A., Bahl, M., Steinberg, L. and Mandel, I. (1987) Col. Surf. 26, 109-121.

G.Rølla
D.Gaare
F.J.Langmyhr
K.Helgeland

Silicon in calculus and its potential role in calculus formation

Faculty of Dentistry, University of Oslo, Geitmyrsveien 71, Oslo 4, Norway

ABSTRACT

This study showed that all the calculus samples analyzed contained silicon. Calculus formed during a period of 5 months contained an average of 132 ppm silicon, whereas older calculus contained up to 2000 ppm. SEM-EDAX analysis indicated that the silicon was present in localized areas. Silicon has been associated with formation of ectopic bone in the urinary system of sheep and cattle. It may be speculated that silicon may play a similar role in formation of dental calculus. An in vitro experiment showed that hydroxyapatite takes up silicic acids. A model explaining some aspects of the calculus- and boneforming potential of silicic acids is suggested.

Recent Advances in the Study of
Dental Calculus

INTRODUCTION

    Silicon is one of the most abundant elements in the
crust of the earth; rocks, minerals, clay and sand contain
silicon as a major constituent. Silicon is able to form an
indefinite number of substances by being connected by oxygen
atoms and forming layers and three-dimensional frameworks.
Silicon has the same role in the inorganic world as carbon has
in the organic world. From soluble rocks silicon goes into the
ground water (as monosilisic acid), which is taken up by
plants and trees, where silicon is included in structural
elements of the plant cells (Fig. 1).
    Silicon is present in high amounts in nails, hair,
connective tissue and enamel, and urine and serum contains
monosilicic acid.
    Silicon is an essential trace element in rats and
chicken. The body weight in these species increases as much as
25-50% on addition of silicon to the food, compared with
control animals on silicon free or nearly silicon free food
(2).
    Silicon induces bone formation in vitro and is presu-
mably also associated with such processes in the animal body.
Silicon appears very early in foci where bone starts to form,
together with calcium. However, phosphorus (as phosphate) soon
takes over as the dominating anion (2).
    Silicon is known to be associated with urolithiasis in
cattle in areas where the ground (and the grass) contains high
amounts of silicon in western USA and Canada and in Australia
(3).
    A few scattered reports indicate that silicon is present
in dental calculus. In the light of the information that
silicon is involved in early bone formation, it also appears
that it may be involved and contribute to the formation of an
ectopic bone like calculus. The aim of the present investigation was to study the silicon content of calculus of known
history, and also of some samples of old calculus deposits.

MATERIALS AND METHODS

    15 samples of calculus which was formed during a period
of 5 months in Indonesian soldiers in Jakarta was collected
and kept in covered plastic vials to avoid contamination with
dust, until analyzed. A further 8 samples were obtained from
older individuals in Oslo and Jakarta which had high amounts

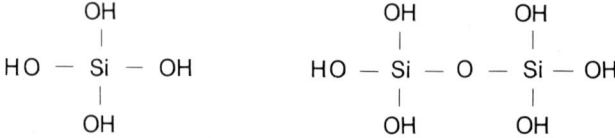

Fig. 1. Monosilicic acid and a condensed silicic acid.

of calculus of unknown history. These samples were collected and kept as described above. The samples were weighed, dissolved in nitric acid and analyzed for silicon by electrothermal atomic absorption spectroscopy. Some samples were also analyzed by inductive coupled plasma atomic emission spectroscopy. One large piece of calculus was invested in plastic, cut with a diamond disc and examined with SEM and EDAX. Uptake of silicon (as silicic acids) by hydroxyapatite was examined by adding 100 mg of Biogel HTP (Bio-Rad) to a series of test tubes each containing 3 ml of diluted and neutralized waterglass (Norsk Medisinaldepot, Oslo) ranging from 500 to 2500 ppm. The suspensions were equilibrated for 1 hour at room temperature, then washed in distilled water 3 times, dried, dissolved in nitric acid and analyzed as described for the calculus samples.

RESULTS

All the calculus samples contained silicon (Tables 1 and 2). The calculus samples which were collected in Jakarta and

Table 1. Chemical analysis of calculus. Samples from young Indonesian individuals, newly formed calculus (5 months).

Ind. no.	Si (ppm)
027	142
130	137
005	66
064	64
045	60
004	83
077	92
085	76
113	199
054	357
100	97
088	181
122	709
055	210
076	532

Table 2. Chemical analysis of calculus. Samples of old, large deposits form elderly individuals in Jakarta and in Oslo.

Ind. no.	Si (ppm)
N 1	154
N2	155
W	101
F	21
Ex	23
1	2220
3/6	283
137	2060

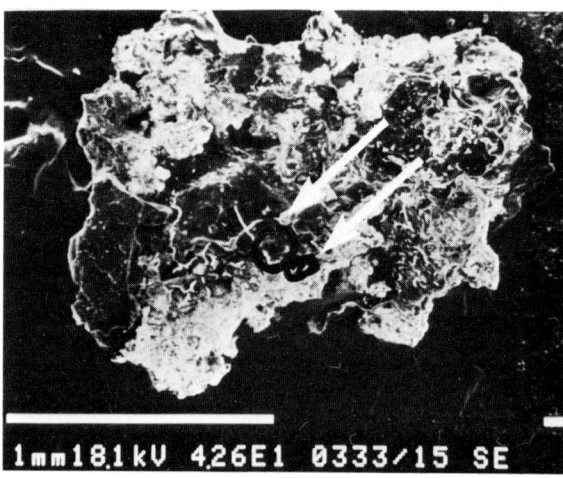

Figure 2. A section of dental calculus seen by SEM. The arrows show areas where silicon was demonstrated by EDAX (Fig.4)

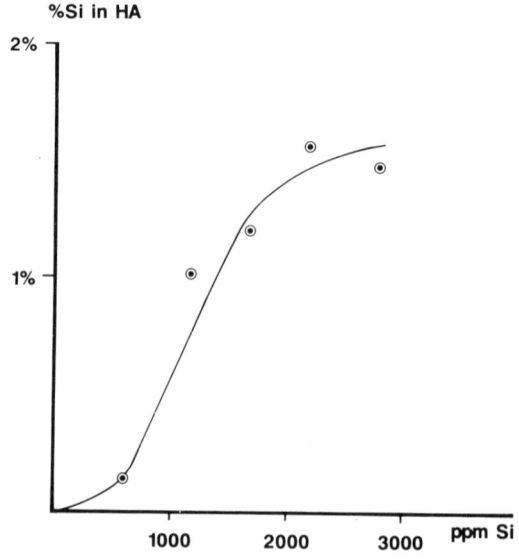

Figure 3. Uptake of Si by hydroxyapatite (HA) 3 ml solution and 100 µg HA

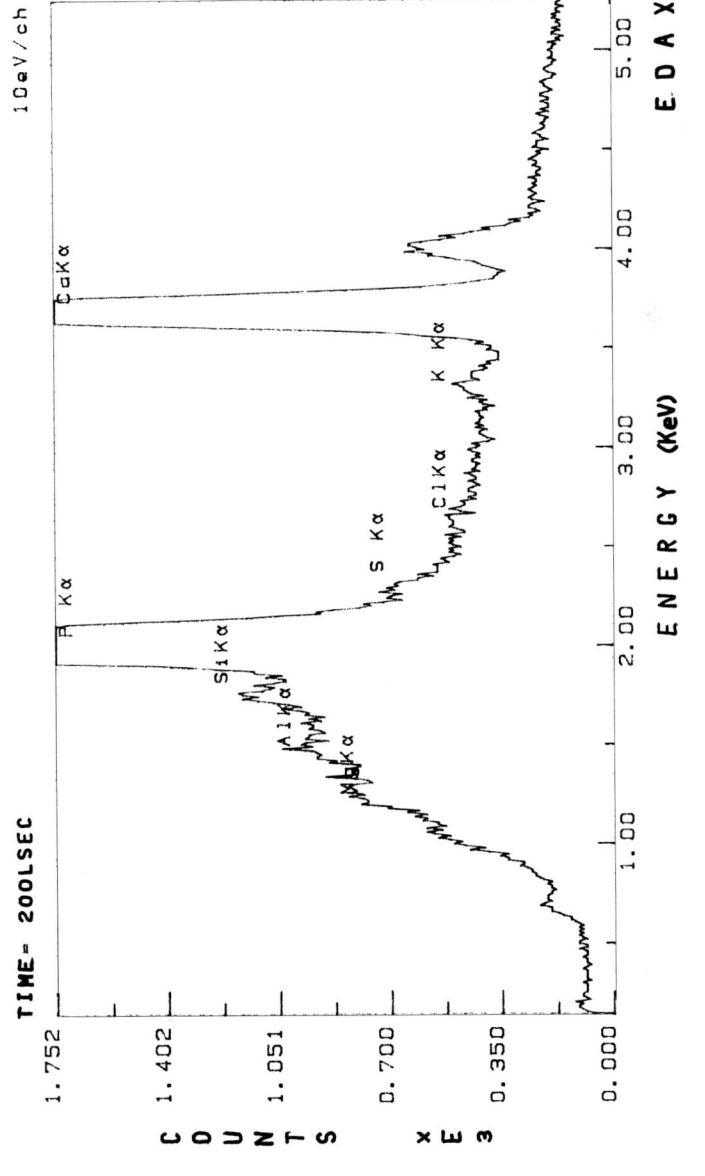

Figure 4. EDAX analysis of an area of a section of dental calculus (arrows Fig.2)

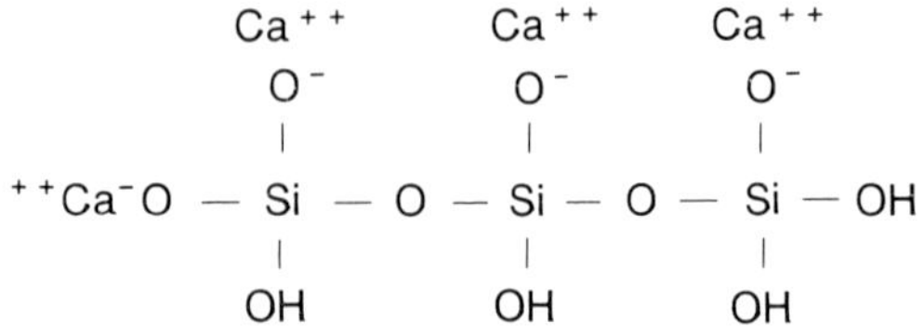

$$\text{Ca}^{++} \qquad \text{Ca}^{++} \qquad \text{Ca}^{++}$$
$$\text{O}^- \qquad\quad \text{O}^- \qquad\quad \text{O}^-$$
$$| \qquad\qquad | \qquad\qquad |$$
$$^{++}\text{Ca}^-\text{O} - \text{Si} - \text{O} - \text{Si} - \text{O} - \text{Si} - \text{OH}$$
$$| \qquad\qquad | \qquad\qquad |$$
$$\text{OH} \qquad\quad \text{OH} \qquad\quad \text{OH}$$

Fig. 5. Calcium binding by a condensed silicic acid. the degree of dissociation of the OH-silanol group increases with increasing pH.

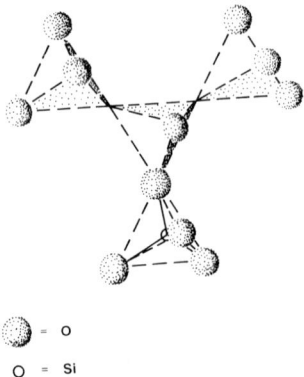

= O

O = Si

Fig. 6. A proposed 3-dimensional structure of a condensed silicic acid. The oxygen atoms are potential calcium binders. This may provide a structural configuration which is favourable for deposition of hydroxyapatite (modified after ref. 7).

which were formed during the last 5 months contained a mean of 132 ppm ranging from 60 to 709 ppm. The variation in the older samples was larger; two samples had values over 2000 ppm Si. The SEM-Edax examinations showed a localized content of silicon (Fig.2). The results of the EDAX is given in Fig. 4 and showed a distinct silicon peak together with calcium and phosphate as expected. Hydroxyapatite took up silicic acids to a maximum of 1.5% (Fig. 3).

DISCUSSION

The study showed that dental calculus always contained silicon, which has also been mentioned in previous reports (4). It thus appears conceivable that silicon may be involved in the mechanism of calculus formation, although the details in this process is not known. The EDAX and SEM study showed that the silicon-containing material was located in specific

areas of the calculus and not evenly distributed. This is the same pattern which has been reported for bone (1,2).
Silicic acid adsorbed to hydroxyapatite. It is known that anions and polyanions bind to the calcium ions on the apatite surface (5) and this is presumably also the mechanism by which silicic acids bind to hydroxyapatite. OH-silanol groups on silica bind cations at neutral and alkaline pH(6), thus we presume that this is the case also in silicic acids (Fig. 5) although their detailed structure is incompletely known. The calcium binding capacity of silicic acids may well be the property which initiates calculus (and bone) formation and which induces precipitation of calcium phosphates in vitro (8). A high local concentration of calcium based on affinity to OH-silanol groups on silicic acids as described above, would attract phosphate ions. The 3-dimensional configuration of the silicic acids may also well contribute to favourable conditions for deposition of hydroxyapatite or other calcium phosphates. The local pH is probably essential because the cation binding by OH-silanol groups is lost at low pH (6).A schematic drawing in Fig. 6 demonstrates this concept.

REFERENCES

1.	Carlisle, Edith M. (1972) Science 178, 619-621.
2.	Carlisle, Edith M. (1982) Nutrition reviews 40, 193-198.
3.	Bailey, C.B. (1967) Am. J. Vet. Res. 28, 1743-1749.
4.	Mc Dougall, W.A. (1985) Archs. Oral Biol. 30, 603-608.
5.	Bernardi, G (1975) Collaques Internationaux du Centre National de la Rescherche Scientific no. 230, 463-65.
6.	Ruvarac, A. (1982) In: Inorganic ion exchange materials (Ed. Clearfield, A) CRC Press, Inc., Florida pp 141-160.
7.	Pauling, L General chemistry, W.H. Freeman and Company, San Fransisco 1948 p 527.
8.	Damen, J. and ten Cate, J.M. (1989). This symposium.

J.J.M.Damen
J.M.ten Cate

Calcium phosphate precipitation is promoted by silicon

Department of Cariology and Endodontology,
Academic Centre for Dentistry Amsterdam,
Louwesweg 1, 1066 EA Amsterdam, The Netherlands

ABSTRACT

To investigate a possible role of silicon in calculus formation we have determined the effects of diluted waterglass on the precipitation of calcium phosphate in vitro. Both spontaneous precipitation and seeded crystal growth from solutions containing 1 mM calcium, 7.5 mM phosphate, 50 mM Hepes, pH 7.2 and 0 - 5 mM silicon were followed by measuring calcium consumption by means of Ca^{++}-selective electrodes. The lag period or induction time which precedes the spontaneous precipitation of calcium phosphate was found to be reduced by 60% in the presence of silicon. By comparing the effects of different waterglass dilutions it was established that polysilicic acid was the active compound and not monosilicic acid. Particulate silica was found to exert a similar stimulating effect as polysilicic acid on spontaneous precipitation. In all cases hydroxyapatite was identified by X-ray diffraction as the predominant crystalline phase. Also the rate of growth of seeded hydroxyapatite crystals was markedly enhanced in the presence of silicon. Brushite seeds grew slowly in absence of silicon, but calcium was rapidly depleted from solution in its presence. X-ray diffraction analysis revealed the brushite seeds to be converted almost completely to hydroxyapatite. Apparently, polysilicic acid acts as a heterogeneous nucleation substrate, favoring the formation of hydroxyapatite over other calcium phosphates.

Recent Advances in the Study of
Dental Calculus

INTRODUCTION

Calcification of dental plaque may either be promoted or inhibited by a variety of compounds of endogenous or exogenous origin. To assess the role of individual components of plaque, a system was developed by which their effects on calcium phospate precipitation in vitro could be determined by measuring the consumption of calcium from reaction mixtures with Ca^{++}-selective electrodes. The experimental set-up allowed ten or more assays to be performed simultaneously, with the data being stored directly on a microcomputer.

As an application of this system, the effect of silicon on calcium phosphate precipitation was studied. Silicon has recently been suggested to promote mineralization of dental plaque: a high calculus formation rate observed in an Indonesian population has been related to the abundance of silicon in the South East Asian diet (1).

As a source of silicon we used waterglass which upon dilution results in the formation of the soluble monosilicic acid $Si(OH)_4$. Silicic acid in turn may condensate to form polysilicic acid or colloidal silica. $Si(OH)_4$ and silica represent most of the silicon found in drinking water and food.

EXPERIMENTAL

Precipitation was determined following the mixing of two buffered solutions which contained calcium and phosphate, respectively. Calcium and phosphate concentrations which have been described to occur in plaque fluid (2) would in absence of inhibitors result in immediate precipitation. Therefore, concentrations were chosen so as to delay precipitation, but the low calcium to phosphate ratio of plaque fluid was maintained.

Into thermostated polypropylene incubation vessels 4 ml 4.0 mM $CaCl_2$/50 mM Hepes, pH 7.2 and 0 - 4 ml diluted waterglass were pipetted, followed by 50 mM Hepes to a total volume of 8 ml. In this mixture Ca^{++}-selective and reference electrodes were equilibrated during 1 hour. The reaction was started by addition of 8 ml 15 mM KH_2PO_4/50 mM Hepes, pH 7.2, which resulted in final concentrations of 1 mM calcium, 7.5 mM phosphate and 0 - 5 mM silicon. All incubations were done at 37 °C under controlled stirring. Reactions were followed by measuring the concentrations of free Ca^{++}.

In crystal growth experiments hydroxyapatite or brushite seeds (specific surface areas 8 and 3 m^2/g, respectively, and kindly donated by Dr.G.S.Ingram) were weighed out in the incubation vessels beforehand so as to give final concentrations of 0.1 mg/ml.

As a parameter for spontaneous precipitation we determined the lag period, or induction time t_i, which occurred before precipitation set in. t_i was defined by the intersection of the horizontal tangent to the Ca^{++} concentration curve following the addition of the phosphate solution, and the tangent to the steepest part of the curve. The chosen conditions resulted in t_i values of about 80 minutes in control incubations.

For X-ray diffraction analysis, precipitates were harvested by filtration, washed with a small amount of ethanol, lyophilized and examined by means of a Philips PW 1327 powder camera with

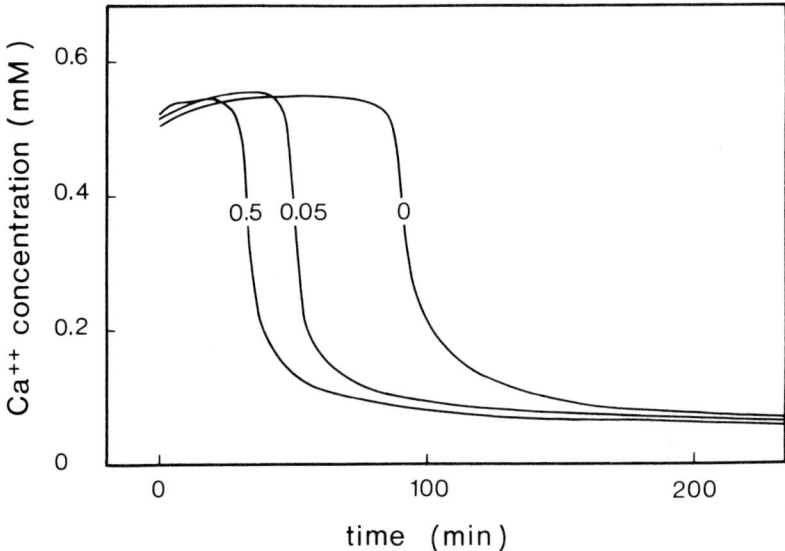

Fig.1   Change in free calcium ions after addition of phosphate
        solution to three incubations which contained 0, 0.05 and
        0.5 mM silicon.

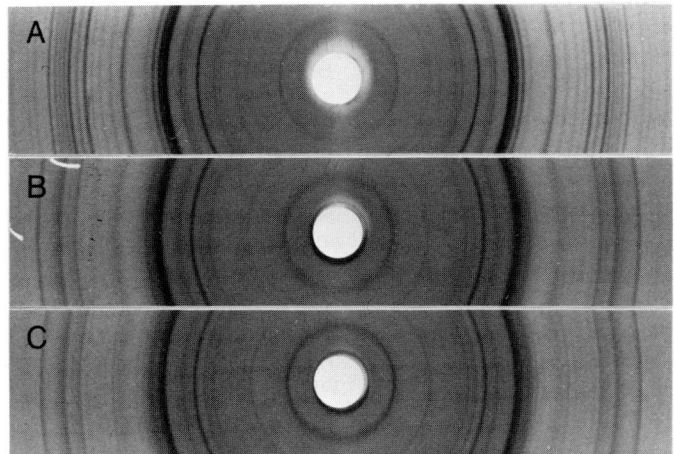

Fig.2   X-ray diffractograms of (A) a commercial hydroxyapatite,
        (B) calcium phosphate precipitated in absence of silicon,
        (C) calcium phosphate precipitated in presence of silicon.

nickel-filtered copper radiation generated at 50 kV and 30 mA anode current.

SILICON STIMULATES HYDROXYAPATITE PRECIPITATION

The effect of silicon on the spontaneous precipitation of calcium phosphate is illustrated in Fig.1 by the change in free Ca^{++} after addition of the phosphate solution to three incubations which contained 0, 0.05 and 0.5 mM silicon, respectively. In the presence of silicon t_i became strongly reduced, with 0.5 mM silicon by even more than 60%, but precipitation rate and final Ca^{++} concentrations seemed not to be affected.

By X-ray diffraction analysis hydroxyapatite was found to be the predominant crystalline phase in all precipitates (Fig.2) The broadened diffraction lines (as compared to those of a commercial hydroxyapatite, Fig.2A) were indicative of a poor crystallinity. No evidence was found of silicon-containing minerals being present in the precipitates, which in view of the low amounts of silicon in the incubations was not unexpected.

Not only spontaneous precipitation of calcium phosphates, but also the growth of seeded crystals was found to be strongly promoted by silicon. Fig.3 shows the initial growth of seeded hydroxyapatite crystals to be enhanced with increasing silicon concentrations. After 4 hours an apparent equilibrium Ca^{++} concentration was reached, which was the same in all incubations.

Seeding with brushite crystals resulted in a different picture (Fig.4): in the presence of silicon the onset of Ca^{++} consumption was delayed, but not in the control incubation. The lag period, which was shorter at the higher silicon concentration, was followed by a rapid depletion of calcium from the solution, which seemed to be independent of the silicon concentration. Although this latter precipitation behaviour resembled the results of the spontaneous precipitation experiment (Fig.1), the lag periods were considerably shorter than t_i values at corresponding silicon concentrations, which indicated the seeds to act as nuclei. After 5 hours apparent equilibrium concentrations were reached, but a distinct difference between the control and the silicon-containing incubations remained, which suggested precipitation of a mineral of lower solubility than brushite in the presence of silicon. This was confirmed by X-ray diffraction analysis. Fig.5A represents the brushite seeds and shows the dotted diffraction lines which are characteristic of a non-rotated sample of large crystals. Following growth in absence of silicon, a similar pattern was found (Fig.5B), but when grown in presence of silicon, brushite had largely disappeared in favor of hydroxyapatite (Fig. 5C). Only a few dots corresponding with the diffraction lines of brushite could still be seen. Apparently, silicon induced the formation of hydroxyapatite on the seeded brushite crystals, whereupon the seeds themselves dissolved as the solution became undersaturated with respect to brushite.

We have compared the effects of silicon on seeded crystal growth with those of fluoride, which is also known to promote the

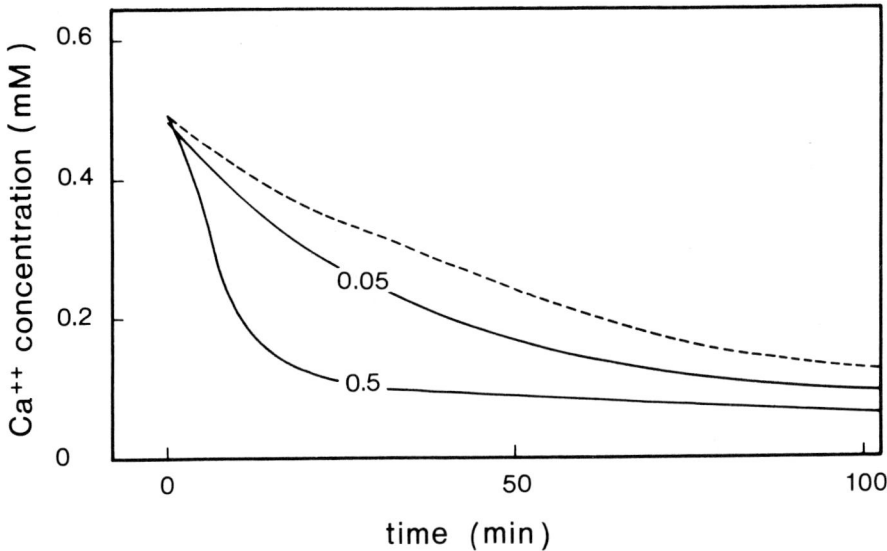

Fig.3   Seeded hydroxyapatite growth in a control incubation
        (----), and in presence of 0.05 and 0.5 mM silicon (———).

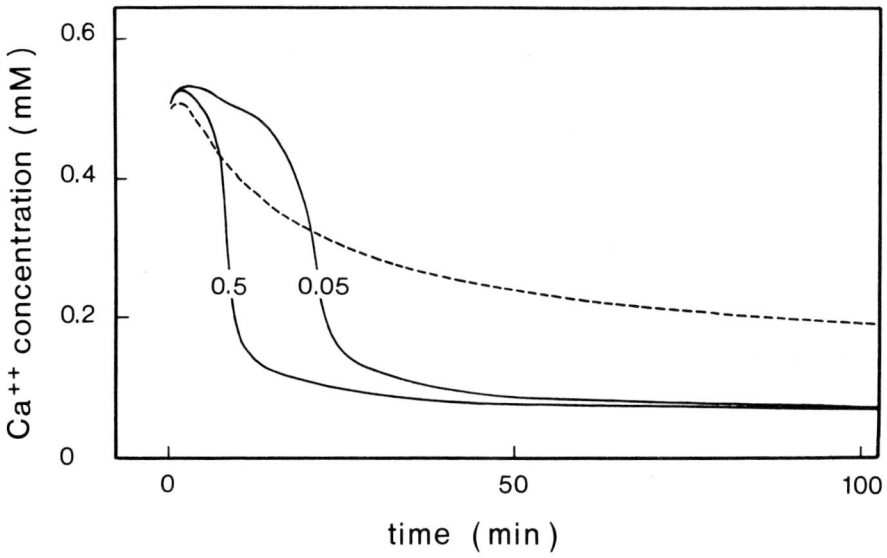

Fig.4   Seeded brushite growth in a control incubation (----), and
        in the presence of 0.05 and 0.5 mM silicon (———).

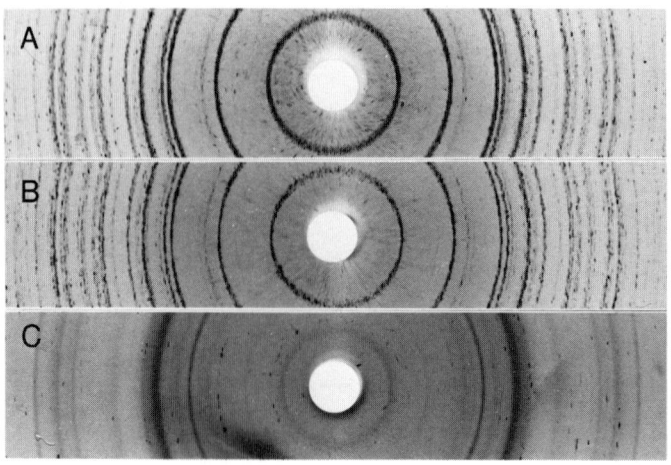

Fig.5 X-ray diffractograms of (A) brushite seeds, (B) brushite grown in absence of silicon, (C) brushite grown in presence of silicon.

formation of apatitic minerals (3). In the presence of 25 or 100 μM KF (about 0.5 and 2 ppm fluoride), the growth of seeded brushite crystals was not affected, and no apatitic mineral was formed (Fig.6). This may be indicative of different mechanisms by which silicon and fluoride promote the precipitation of calcium apatites. In contrast, the initial growth of seeded hydroxyapatite crystals was accelerated by both silicon and fluoride (Fig.7). No addition of these effects could be observed when silicon and fluoride were simultaneously included in the precipitation mixture. It cannot be excluded that such a cumulative effect may occur under appropriate experimental conditions. A synergistic action of silicon and fluoride on the mineralization of bovine achilles tendon collagen in vitro has recently been reported (4).

PRECIPITATION-STIMULATING ACTIVITY IS ASSOCIATED WITH POLYSILICIC ACID

We have made some effort to identify the nature of the silicon compound responsible for the stimulation of hydroxyapatite precipitation.
Dilution of waterglass at physiological pH results in the formation of monosilicic acid, which in turn may condensate to form polysilicic acid. This autopolymerization is concentration-dependent. The concentration above which a monosilicic acid solution becomes unstable has been reported 3.5 mM by one author and 2 mM or even 0.1 mM by others (cited in Ref.5). The preceding results were obtained by using a stock solution of 20 mM silicon. It is very likely that in this stock solution polysilicic acid had formed. In fact, there were some indications for such polymerization to occur. For example, we obtained very irreproducible results with freshly-made waterglass dilutions. Consistent re-

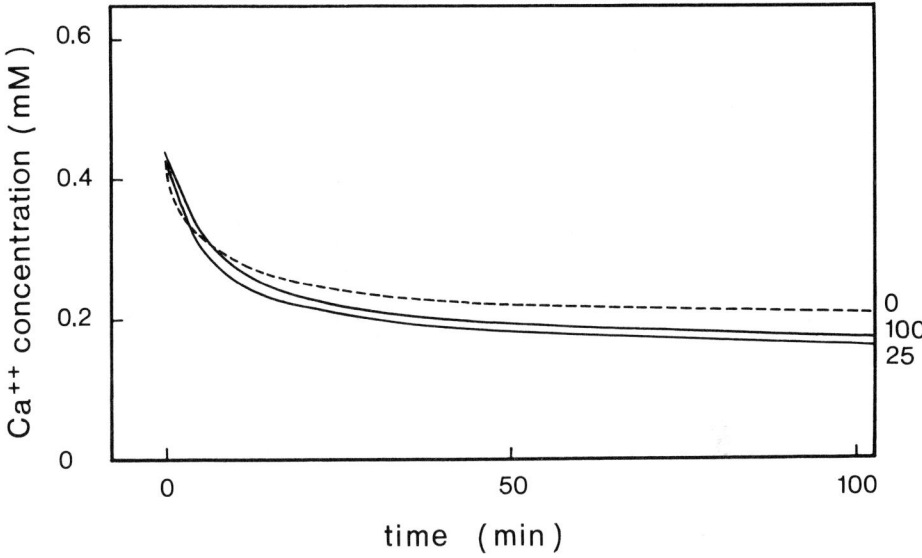

Fig.6 Seeded brushite growth in a control incubation (----), and in the presence of 25 and 100 μM KF (——).

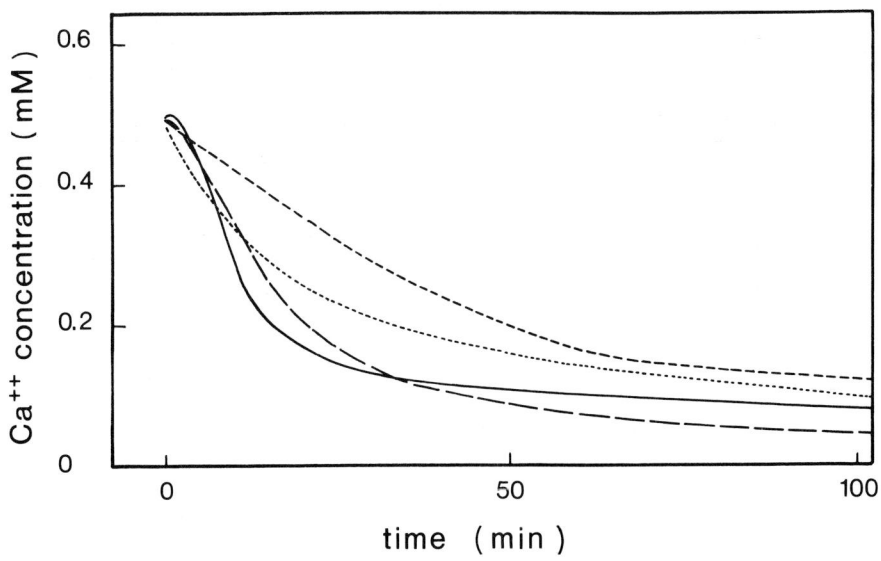

Fig.7 Seeded hydroxyapatite growth in a control incubation (----); in the presence of 0.5 mM silicon (——); 25 μM KF (·········); 0.5 mM silicon + 25 μM KF (— —).

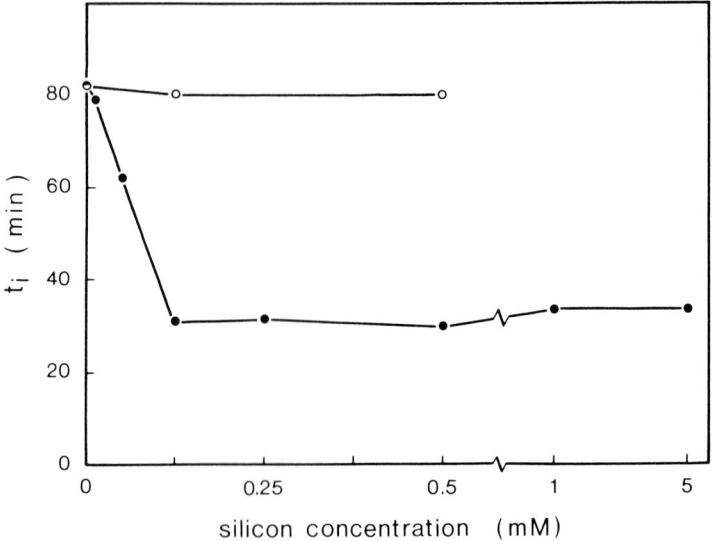

Fig.8   Effect of silicon on the induction period t_i of
        spontaneous precipitation of calcium phosphate. Silicon
        was added as (o) 2 mM or (●) 20 mM solution.

sults were obtained only after some time, and therefore, we
always used at least one-day old solutions; these one-day old
solutions sometimes were slightly opalescent.
    In order to discriminate between possibly different effects of
mono- and polysilicic acid, we have compared stock solutions of
20 and 2 mM silicon in which polymerization may and may not
occur, respectively. The effects of both solutions on spontaneous
precipitation are shown in Fig.8. When various amounts of the 2
mM solution were added to the incubations, no changes in t_i were
found. When instead the concentrated silicon solution was used,
t_i gradually decreased with an increasing silicon concentration
in the incubations, until at 0.15 mM a maximal reduction was
reached. No further changes in t_i could be observed when more
silicon was included in the incubations up to 5 mM. These results
indicated that the capability of a diluted waterglass solution to
stimulate calcium phosphate precipitation may be attributed to
the presence of polysilicic acid.
    Association of precipitation-stimulating activity with the
polymer could be affirmed by gel filtration of the 20 mM stock
solution on Sephadex G-25: precipitation-stimulating activity
eluted exclusively in the void volume fractions, which indicated
a particle weight much higher than the molecular weight of monosilicic acid.
    When polysilicic acid is diluted, it is known to depolymerize
again. Dilution of the 20 mM silicon solution resulted in a slow
diminution of its activity, but it took several days before this
became apparent. It explains why in our experiments activity

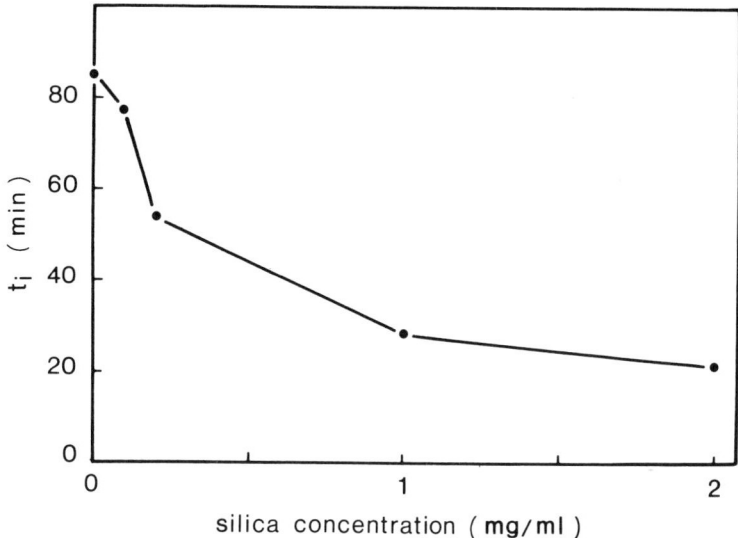

Fig.9   Effect of particulate silica on t_i of spontaneous
        precipitation of calcium phosphate.

could still be measured, when samples of this solution were
diluted by their addition to the incubation mixtures.

As the condensation of silicic acid proceeds, larger particles
develop which are known as silica. In many plants, excess silicon
taken up as monosilicic acid is deposited as silica (6). We have
tested an industrial silica product for its capability to stimulate spontaneous precipitation. Therefore, various amounts of a
suspension of silica particles (70-230 mesh; Merck, Darmstadt,
F.R.G.) were added to precipitation mixtures instead of diluted
waterglass. Fig.9 shows that also this particulate silica promoted the precipitation of calcium phosphate by reducing the induction time. When compared with the waterglass-derived solution,
this reduction was of similar magnitude (from 80 to 25 minutes),
but the required amount of silicon was 100 times larger. This
probably reflected the substantial difference in size and surface
area between the large silica particles and the polymers in the
waterglass-derived solution.

Polysilicic acid and silica probably act as heterogeneous nucleation substrates, which stabilize developing hydroxyapatite
nuclei and thus reduce the time which is required for the formation of nucei of stable size (critical nuclei) (7). In case of
crystal growth, polysilicic acid may in a similar way stabilize
secundary nuclei on the surfaces of both hydroxyapatite and
brushite seeds. The induction of hydroxyapatite precipitation on
brushite crystals indicates that polysilicic acid favors formation of hydroxyapatite nuclei over other calcium phosphate
phases.

CONCLUSIONS

The present results expose silicon as a possible promotor of calculus formation. Silicon was found to stimulate the precipitation of hydroxyapatite under conditions which more or less resemble those in plaque fluid. The active silicon compound appeared to be polysilicic acid and not monosilicic acid, and large silica particles showed similar effects as polysilicic acid.

These results are consistent with the observations made by EDAX of silicon in calculus to occur in foci rather than evenly distributed (8,9). It is not unlikely that such inclusions represent silica particles, for example derived from agricultural food products such as rice, which were trapped in dental plaque and there acted as nuclation substrates.

REFERENCES

1. Gaare, D., Rølla, G. and Van der Ouderaa, F. (1989) this symposium.
2. Tatevossian, A. (1987) Archs. Oral Biol. 32, 201-205.
3. Moreno, E.C. and Varughese, K. (1981) J. Crystal Growth 53, 20-30.
4. Anasuya, A. and Narasinga Rao, B.S. (1983) Biochem. Med. 30, 146-156.
5. Ciba Found. Symp. 121 (1986) Silicon Biochemistry, John Wiley, Chichester, UK.
6. Sangster, A.G. and Hodson, M.J., Op. Cit. (Ref. 5) pp.90-103.
7. Nancollas, G.H. (1966) Interactions in electrolyte solutions, Elsevier, Amsterdam, NL.
8. McDougall, W.A. (1985) Archs. Oral Biol. 30, 603-608.
9. Rølla, G., Gaare, D., Langmyhr, F.J. and Helgeland, K. (1989) this symposium.

D.Gaare[1]
G.Rølla[1]
F.van der Ouderaa[2]

Comparison of the rate of formation of supragingival calculus in an Asian and a European population

[1]Faculty of Dentistry, University of Oslo, Geitmyrsveien 71, Oslo 4, Norway and [2]Unilever Research, Port Sunlight Laboratory, Bebington, Merseyside L63 3JW, UK

ABSTRACT

A comparison of formation rate of calculus was performed between 74 male students in Oslo and 77 soldiers in Jakarta of the same age. It was found that the formation rate was higher in the Asian population and that more teeth acquired calculus in this group. The difference in formation rate was not caused by age, sex, toothbrushing frequency or toothpaste abrasivity. It is suggesting that differences in eating habits and food may account for the difference. In particular consumption of rice which contains silicon and which has a low abrasivity may be a significant factor in the high formation rate in the Asian group.

Recent Advances in the Study of
Dental Calculus

D.Gaare, G.Rølla and F.van den Ouderaa

___

South East Asian populations exhibit high amounts of calculus (1,2). The location of calculus, the amounts present and its hardness are different from what is usually seen in European or American populations. Large deposits cover the lingual surfaces of all the teeth and are also found buccally on many locations in addition to the reported "normal" locations in Europeans, i.e. on the first permanent molar in the upper jaw. It is not known whether the high amounts of calculus in Asian populations are caused by a high rate of formation or whether they represent life-long accumulations of calculus.

The aim of the present study was to compare the rate of formation of calculus in comparable populations under as standardised conditions as possible in Oslo, Norway and Jakarta, Indonesia.

MATERIAL AND METHODS

The Norwegian test panel consisted of 74 male students aged 22-26 years while the Indonesian panel consisted of 77 professional soldiers of the same age.

All were given motivation for oral hygiene and were individually instructed in toothbrushing. Plaque was demonstrated with a disclosing agent and the limitations of their brushing efficiency was pointed out to members of both panels at 3 subsequent appointments.

A thorough prophylaxis, including removal of sub- and supragingival calculus with scalers and ultrasonic equipment was then carried out in dental clinics equipped with good light and compressed air. Virtually all the Indonesians had high amounts of calculus and it took an experienced clinician a mean of 1 hour per person to remove it. A more detailed description of the Indonesian population has been given in a recent publication (3). Only about one third of the Norwegians had calculus at the start of the study, and this was much easier to remove. The teeth were extensively dried during the final check for calculus removal. Both test panels were supplied with new toothbrushes and with toothpastes with the same abrasivity (alumina).

Calculus in the Norwegian panel was scored after 6 months, and in the Indonesian panel after 5 months. Scoring was according to the V-M index (4). For both panels the scoring was done by the same clinician (D.G.). The teeth were pumiced with a rotating rubber cup and dried extensively before the calculus index was measured with a graded probe. Only calculus that was firmly bound to teeth was thus recorded. The number of bleeding points (5) on three surfaces of 28 teeth, in both groups, were recorded at the end of the study.

Very few of the Norwegian students smoked whereas a majority of the Indonesians smoked 1-2 cigarettes per day.

RESULTS

The number of buccal and lingual surfaces with calculus scores from 0-3 is given in Table 1. The Indonesian soldiers had markedly more calculus than the Norwegian students, in

___

	Indonesians					Norwegians				
V-M scores	0	0.5	1	2	3	0	0.5	1	2	3
Upper jaw										
Buccal surfaces	990	61	15	6	0	1024	2	0	0	0
Lingual surfaces	1071	2	0	0	0	1026	0	0	0	0
Lower jaw										
Buccal surfaces	1050	7	0	0	0	1029	1	1	0	0
Lingual surfaces	697	262	83	12	4	906	98	20	4	1

Table 1. Calculus formed during the experimental period (V-M scores) in 77 Indonesian soldiers and 74 Norwegian students under standardized conditions.

spite of the experimental period being one month longer in Norway. On the maxillar buccal surfaces, 82 Indonesian teeth had scores of 0.5-3, whereas only 2 teeth with a calculus score of 0.5 were found in the Norwegian panel.

On the lingual surfaces in the lower jaw, the total number of teeth with calculus was 361 in the Indonesian panel and 123 in the Norwegian. The distribution of calculus on the individual teeth is given in Tables 2 and 3. In addition to having calculus on the buccal surfaces of the two permanent molars in the upper jaw, the Indonesians had calculus on all the lingual surfaces in the lower jaw. In particular, the lingual surfaces of the premolars commonly exhibited calculus. The Norwegian group rarely showed calculus beyond the 6 lower incisors. The distribution of calculus was asymmetrical in the mouths of both populations (Fig.1). An interesting observation was that the only Norwegian student who showed very high calculus scores was a South Korean. This student alone accounted for all the score 2 and 3 in the Norwegian group. The student had lived in Oslo for 2 years and had maintained his Asian diet, as far as possible, with rice being a major carbohydrate source.

The frequency of toothbrushing was examined by means of a questionnaire. The Indonesians brushed an average of 3 times per day whereas the Norwegians brushed 2 times per day. There was no significant correlation between amounts of calculus present at the end of the test period and the number of bleeding points in any of the groups (Table 4).

It was often observed during the clinical examination of older Indonesians (3) that large amounts of ^R calculus could be present on only one side of the dentition. This happened when the chewing was performed on one side due

CALCULUS DISTRIBUTION AMONG INDONESIANS AND NORWEGIANS

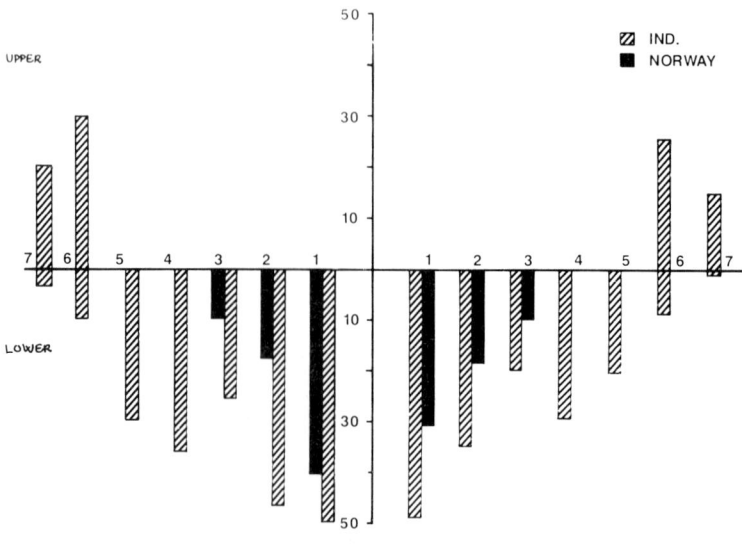

Figure 1. The figure shows distribution of calculus on individual teeth (all surfaces) in Indonesians and Norwegians.

to a painful tooth, and the calculus was found on the non-functional side.

DISCUSSION

The study showed that the rate of calculus formation was higher in the Indonesian population tested than in the Norwegian. The location of the calculus in the two groups was also different; calculus on the lingual aspects of the premolars and molars in the lower jaw was observed almost solely in the Asian group. Calculus on the buccal surfaces of the molars in the upper jaw was also a characteristic of this group. Calculus on these surfaces is certainly seen in Norwegians in general, but it obviously takes more time to form than the 6 months of the present study.

Age, sex, abrasivity of the toothpaste, toothbrushing frequency and method cannot explain the differences between the Indonesian and Norwegian groups. The standard of oral hygiene does not depend entirely on frequency of toothbrushing, but since no relationship was seen between the number of bleeding points and the rate of calculus formation, the differences between the groups were probably not due to a difference in oral hygiene standards. However, both groups showed less calculus on the left side than on the right side of the mouth (probably because the left side is more accessible to right-handed people and is therefore cleaned more effectively). This suggests that rate of calculus formation can be reduced to some extent by toothbrushing. Results obtained by Schroeder (6) do not suggest such an asymmetrical distribution of calculus, but these were results from a study of calculus formation on plastic foils and no brushing had been allowed.

Thus, the reason for the greater rate of formation of calculus in the Indonesian group must lie elsewhere. Differences in food, eating habits and mineral content of the drinking water of the two test groups may provide this reason. High consumption of sucrose and frequent eating between meals will cause plaque pH to fall. Since calcium phosphates are deposited more easily at high pH, sucrose consumption can probably be related to rate of calculus formation. The relatively low sucrose consumption of the Indonesians (10 kg per person per year) may thus be an important factor (3). The rice-based diet, with low sucrose content, will presumably provide an alkaline environment in the mouth - a condition which encourages precipitation of calcium phosphate from a saturated solution like saliva, as indicated above (7). Less frequent eating between meals by the Indonesian group may contribute to this situation. The drinking water in Oslo is very soft, whereas the water in Jakarta is hard. Water hardness may thus also influence the rate of calculus formation.

The relatively low abrasivity of the rice diet may encourage calculus build-up. Furthermore, since the rice is steamed, the minerals in the hard water will be concentrated in the rice which is itself rich in some minerals.

The observation that the one participant in the Norwegian group who had an Asian background formed markedly more calculus than the others is supported by a general impression gained by dental practitioners in Norway, namely that the guest workers from South East Asia generally form calculus more rapidly than the rest of the population. This point, however, has not been specifically investigated. These observations suggest that the food is important, because the immigrants tend to retain their traditional eating habits, whereas the drinking water necessarily is similar to that of the Norwegian population. It has recently been suggested that silicon, which is abundant in rice, may be one factor involved in the high rate of calculus formation in Asian populations. Silicon is known to induce bone formation under certain circumstances, and this includes ectopic bone formation (for review see 8). The observations that dental calculus contains silicon, that silicon increases the rate of calculus formation in rats and that it induces precipitation of calcium phosphates in vitro, as shown in other papers in this symposium (9, 10, 11) supports this contention.

Table 2. Distribution of calculus according to the V-M index on individual teeth in 77 Indonesian soldiers.

		Right							Left						
		7	6	5	4	3	2	1	1	2	3	4	5	6	7
BUCCAL (upper jaw)															
surface	0	64	44	76	77	77	77	77	77	77	77	77	76	52	62
	05	12	20											18	11
	1	1	6											5	4
	2		5											1	
	3														
Missing			2	1									1	1	
(lower jaw)	0	76	70	76	77	76	77	74	75	76	76	77	75	68	77
	05		1			1		1	2	1	1				
	1														
	2														
	3														
Missing		1	6	1				2					2	9	
LINGUAL (upper jaw)															
surface	0	77	75	76	77	77	77	77	77	77	77	77	76	75	76
	05														
	1													1	1
	2														
	3														
Missing			2	1									1	1	
(lower jaw)	0	74	61	47	40	51	30	25	29	43	58	49	54	62	74
	05		8	20	28	22	36	26	28	29	16	22	19	5	3
	1	1	2	9	9	4	10	20	16	4	3	4	1		
	2	2						3	3	1		1	2		
	3						1	1	1		1				
Missing			6	1				2					2	9	

The asymmetrical accumulation of calculus in individuals with impaired unilateral chewing function of the dentition was concentrated on the nonfunctional side. The pellicle contains substances which inhibits deposition of calculus on teeth (12). It appears that unilateral dysfunction may cause a slow turnover of pellicle formation, due to low friction between the food and the teeth, and thus to loss of inhibition of the pellicle on the non-functional side. Long-term colonization of the pellicle by bacteria, or proteolytic activity in saliva or plaque fluid (13), may cause denaturation of phosphoproteins in the pellicle and loss of inhibiting activity. This is probably an important aspect of calculus formation because the presence of tooth surfaces for nucleation is necessary to obtain the firm binding of calculus to the teeth which is seen in the clinic (14).

Table 3. Distribution of calculus according to the V-M index on individual teeth in 74 Norwegian students.

	\<Right\> 7	6	5	4	3	2	1	\<Left\> 1	2	3	4	5	6	7
BUCCAL (upper jaw)														
surface 0	74	74	74	70	74	74	73	74	74	74	69	74	73	73
05													1	1
1														
2														
3														
missing				4			1				5			
(lower jaw) 0	74	74	74	72	74	73	74	73	74	74	72	74	73	74
05						1								
1								1						
2														
3														
missing				2							2	1		
LINGUAL (upper jaw)														
surface 0	74	74	74	70	74	74	73	74	74	74	69	74	74	74
05														
1														
2														
3														
missing				4			1				5			
(lower jaw) 0	74	74	74	71	65	55	41	41	55	64	71	74	73	74
05				1	8	16	23	25	15	9	1			
1					1	2	8	7	2					
2							1	1	1	1				
3							1							
missing				2		1		1			2	1		

Table 4. Correlation between bleeding points and calculus scores

	Indonesian	Norwegian
Average BP v	0.227	0.202
Average whole mouth calculus P=	0.052	0.099
Average BP v	0.223	0.208
Average Volpe-Manhold calculus P=	0.057	0.088

REFERENCES

1. Wei, S.H.Y, Yang S, Barnes, D.E. (1986) Comm. Dent. Oral Epid. 14, 19-23.
2. Løe, H., Anerud, H., Boysen, H. and Morrison E. (1986) J. Clin. Periodontol. 13, 431-40.
3. Gaare, D., Joelimar, F.A., Rølla, G. ana Ouderaa. F.v.d. (1989) Scand. J. Dent. Res. In press.
4. Volpe, A.R., Manhold, J.H., Hazen, S.P. (1965) J.

Periodontol. <u>36</u>, 292-98.
5.   Ainamo, J. and Bay, I. (1975) Int. Dent. J. <u>25</u>, 229-235.
6.   Schröder, H.E. (1969) Han Huber, Bern
7.   Driessens, F.C.M. (1982) Karger, Basel 154-58.
8.   Carlisle, E.M. (1982) Nutrition Reviews <u>40</u>, 193-198.
9.   Rølla, G., Gaare, D., Langmyhr, F.J. and Helgeland, K. (1989) This symposium.
10.   Damen, J. and ten Cate, J.M. (1989) This symposium.
11.   Guggenheim, B. and Rølla G. (1989) This symposium.
12.   Hay, D.I. and Moreno, E. (1979) Eds. I. Kleinberg, S.A. Ellison and I.B. Mandel. IRL Press, New York 45-58.
13.   Wanatabe, T. and Morita, M. (1989) This symposium.
14.   Busscher, H.J., Uyen, H.M., van Dijk, L.Z. (1989) This symposium.

G.Rølla[1]
B.Guggenheim[2]
R.Schmid[2]

The effect of silicon on dental calculus formation in the rat

[1]Dental Faculty, University of Oslo, Oslo 4, Norway
and [2]Animal Research Unit, Dental Institute, University
of Zürich, Plattenstr. 11, CH-8028 Zürich, Switzerland

ABSTRACT

The experiments showed that silica (insoluble) and magnesium
trisilicate (slightly soluble) both could induce dental
calculus in the rat. Silica induced increased rate of
calculus formation when added to the food. This effect was
visible after 35 days, but not after 49 days. Magnesium
trisilicate showed no effect when added to the food. However,
applied by stomac tube it induced increased calculus formation after 49 days. Silica also showed effect after 49 days
applied in the same way. It is suggested that silica may have
a local effect by being retained in the plaque, and having a
cation-binding potential, presumably causing a high local
concentration of calcium which may be conducive to calculus
formation. Silicon from magnesium trisilicate probably
affects calculus formation systemically through inducing high
concentrations of silicic acids in saliva. Silicic acids are
also calcium binders and may operate through the same
mechanism as solid silica in plaque.

Recent Advances in the Study of
Dental Calculus

INTRODUCTION

It is well known that silicon can be involved in the formation of bone both in vitro and in vivo and reports are
available which indicate that silicon also can be associated
with formation of ectopic bone; urinary calculi are formed in
sheep and cattle where the soil and the grass are rich in
this element (1). It has also been suggested that silicon can
be involved in the mechanism of dental calculus formation and
that this can be an aspect of the observed high calculus
formation rate in Asian populations (2,3) which has the
silicon-rich rice as a major diet component. This rice is
also in many cases steamed in silicon-rich water in vulcanic
areas.
The aim of the present study was to investigate the possible
role of silicon on dental calculus formation in rats.
Silicon, as silica or silicate, was added to the diet or
applied through a stomac tube.

MATERIALS AND METHODS

The study was conducted on 10 litters of OM-rats for an
experimental time of 35 days and 8 litters for an experimental time of 49 days respectively, each litter consisting of 6
animals. On day 20, littermates were distributed at random
among the following treatments: 1) Control, diet 2000fs; 2)
Diet 2000fs containing 2% Aerosil; 3) Diet 2000fs containing
2% magnesium trisilicate; 4) Control, diet 2000fs and
intubation of 2 ml H_2O once daily; 5) Diet 2000fs and
intubation of 200 mg Aerosil once daily; 6) Diet 2000fs and
intubation of 200 mg magnesium trisilicate once daily. The
basis diet 2000fs contained: 63% wheat fluor, 28% skim milk
powder, 5% brewer's yeast, 2% protein-vitamin-mineral
supplement (Gevral Protein[R], Lederle Laboratories, Great
Britain), 1% sodium chloride and 1% powdered sucrose. All
ingredients of the diet were finely powdered and mixed
thoroughly. Diet and tap water were available ad libitum. The
rats were weighed at the beginning and end of the experiment.
At the completion of the study the rats were anesthetized and
decapitated. The 1st and 2nd maxillary molars were scored for
pronounced macroscopic differences in the amount of calculus
deposits. For this, the maxillary molars were inspected with
the aid of a dissection microscope at magnification 32x for
the location and extent of hard calculus deposits. Data were
recorded blind by on charts. All surfaces of the right and
left 1st and 2nd maxillary molars were divided into 20
scoring areas, and each area covered with detectable hard
deposits was assigned one scoring unit. Sources of silicon
were Aerosil 300 (SiO_2) (Degussa AG, Germany) and magnesium
trisilicate ($3MgO_2$ $3SiO_2$ nH_2O) (Serva, Germany). To investigate any direct effect, diet 2000fs mixed with either 2% of
Aerosil or magnesium trisilicate was used. Consumption of
diet 2000fs was approximately 10 g per animal and day during
the experimental period, which represents an intake of 200 mg
of silicon-containing compounds per day. To test the effect

Tabel 1

Average amount of calculus (CU) per rat. (N=10 in testperiod 35 days and N=8 in testperiod 49 days. Weight gains are given in grams).

Method	After 35 days		After 49 days	
	CU	g	CU	g
-1 Control	5.0+3.8	166	15.6+4.1	198
-2 Aerosil 300 (SiO_2) 2% in the diet	9.4+6.6*	165	16.0+2.8^{ns}	203
-3 Magnesium trisilicate 2% in the diet	4.4+3.6^{ns}	155	15.6+3.8^{ns}	193
-4 Control gastric intubation of water, 2ml once a day	10.2+4.5	169	13.8+3.6	204
-5 Aerosil 300, gastric intubation 200mg once a day	12.3+4.9^{ns}	165	19.4+0.9**	202
-6 Magnesium trisilicate gastric intubation 200 mg once a day	10.9+4.5^{ns}	171	18.5+1.7**	208

* $P<0.05$ ** $P<0.01$ ns   not significant

of a possible systemic effect by silicon in calculus formation in rats (through saliva) 200 mg of Aerosil or magnesium
trisilicate suspended in distilled water was applied daily at
4h p.m. by gastric intubation.

RESULTS

Calculus formation after 35 days

The diet containing silicon in the form of Aerosil (silica)
increased the calculus formation ($P<0.05$) compared with the
control group. No such effect was seen in the group which
received magnesium trisilicate.
Suspensions of Aerosil or magnesium trisilicate applied by
stomac tube as explained above did not increase the calculus
formation compared with the control rats.

It was observed that Diet 2000f with 1% sucrose caused a calculus formation of 5 units in the control group, whereas the rats which received distilled water through stomac tubes exhibited a calculus formation of 10.2 units. This difference is statistically significant ($P<0.01$).

Calculus formation after 49 days

The calculus formation was higher in all groups after 49 days than after 35 days.
The groups which received silicon added to the diet showed no increased calculus formation after 49 days. However, application of silicon by stomac tube increased calculus formation in the rats significantly both in those receiving Aerosil and those receiving magnesium trisilicate.
The calculus formation was higher in the control groups which received Diet 2000f compared with the group which in addition received distilled water through stomac tubes.
The dental calculus which was formed during presence of silicon appeared to have a different morphology. In contrast to the calculus formed in the absence of silicon it was thinner and had a glasslike transparent appearance.

DISCUSSION

Silica (Aerosil 300) is mainly insoluble whereas magnesium trisilicate is soluble in the stomac (at low pH). Silica is known to exhibit OH-silanol groups on the outer surface which bind cations at neutral- and acid pH (4). It thus appears likely that the silica in the food could be retained in plaque and bind calcium which could trigger or increase the rate of calculus formation. Magnesium trisilicate has no such effect and would thus not be expected to have any local effect. The effect by silica in the food was seen after 35 days but not after 49 days, can be caused by the well known fact that a maximum amount of calculus is reached after a certain time; a higher rate of calculus formation can thus not be demonstrated at this stage.
Magnesium trisilicate induced higher calculus after 49 days when applied through stomac tube. This effect is probably systemic in nature. The effect by silica applied by stomac tube may be due to a certain solubilization in the alkaline part of the digestive system and thus by a systemic mechanism, or that some insoluble particles of silica found its way into the oral cavity after the application with stomac tube.

REFERENCES

1.   Carlisle EM. (1972) Science 178, 619-21.
2.   Gaare D, Rølla G, v.d.Ouderaa F. (1989) This symposium.
3.   Gaare D, Joelimar FA, Rølla G, v.d.Ouderaa F. (1989) Scand J Dent Res (in press).
4.   Ruvarac A. In: Inorganic ion Exchange materials (ed. A Clearfield) (1982) (RC Press Inc. Florida) pp 141-160.

SESSION IV.
Calculus Inhibition

E.C.Moreno[1]
T.Aoba[1]
A.Gaffar[2]

Physical chemistry of calculus formation

[1]Forsyth Dental Center, 140 Fenway, Boston, MA 02115, USA and [2]Colgate-Palmolive Research Center, 909 River Road, Piscataway, NJ 08854-5596, USA

ABSTRACT

Formation of supragingival dental calculus occurs in a micro-environment limited by the plaque-saliva and plaque-tooth inter-faces. Depending on the proximity to these interfaces, the mineralization process will be affected, to some extent, by the driving forces for precipitation existing in saliva and the plaque fluid, PF; these forces are the degree of saturation, DS, of these fluids with respect to various calcium phosphates. In addition, the presence of inhibitors affects the mineraliza-tion kinetics. Supersaturated solutions were prepared according to the DS values of saliva and PF to study the effect of endo-genous (proline-rich salivary proteins, PRP) and exogenous (phytate, pyrophosphate, polyvinyl phosphonate) inhibitors of crystal growth. In all cases, the inhibitory activity was directly related to adsorption coverages; the latter were calcu-lated with parameters obtained from the adsorption isotherms of the various inhibitors. Inhibitors of low molecular mass, at optimal concentrations, provided complete inhibition of crystal growth even when the precipitation driving force was increased substantially by the addition of F^- to the supersaturated solution; at lower concentrations, these inhibitors exhibited induction periods after which crystal growth proceeded at rates comparable to those in systems without inhibitor. Inhibitors with relatively bulky molecular masses, did not provide complete inhibition upon the addition of F^-, indicating the presence of "hidden" crystal growth sites even at maximal coverages. An important finding was that there is some inhibitory specificity depending on the nature of the seeds. Thus, pyrophosphate is a much better inhibitor than phytate for apatitic growth; however, for the seeded growth of dicalcium phosphate dihydrate, the effectiveness of the two inhibitors is completely reversed.

Recent Advances in the Study of
Dental Calculus

INTRODUCTION

The ultimate step in any biomineralization process, i.e., the deposition of mineral, must conform to the driving forces operating on the system. In addition, kinetics of the process may be affected by the presence of specific constituents of the medium in which the mineral is formed. Consequently, in advancing mineralization models, at a minimum, it is necessary to define 1) the degree of saturation (DS) of the biological medium with respect to all possible precipitating phases, 2) the processes by which accretion of mineral is maintained (e.g. active or passive transport of components and crystal growth), and 3) the possible control of deposition rates by the presence of crystal growth inhibitors or accelerators. The rule in biological systems is to have a variety of interfaces and surfaces which act as nucleators; for this reason, nucleation is not considered here to be a limiting step in biomineralization.

The determination of DS values of the medium will indicate which phases may or may not form for thermodynamic reasons. The processes leading to the accretion of mineral may involve a very direct cellular participation, e.g., formation of extracellular matrices, non-resorptive as in osteogenesis and resorptive as in amelogenesis, or no active cellular participation at all; the latter is the case of ectopic calcification including the formation of dental calculus.

Dental calculus forms in a micro-environment located between plaque-saliva and plaque-tooth interfaces. Therefore, it is reasonable to assume that mineralization taking place towards the plaque-saliva interface will be affected, to some extent, by the properties of the saliva. Similarly, calculus formation in the vicinity of the tooth surface will be determined, to a large extent, by the properties of the plaque fluid. The presence of crystal growth inhibitors (proteins and peptides) in saliva has been known for a considerable time (1-4); their biological function is to keep the saliva supersaturated with respect to the enamel mineral thus preventing enamel losses by dissolution. Also, these inhibitors prevent mineral accretion on the tooth surfaces which would compromise dental morphology. There are also important exogenous inhibitors of crystal growth, e.g., pyrophosphate and other condensed phosphates, that have practical application in the formulation on dentifrices or mouthwashes.

In the present article, laboratory models of mineralization applicable to calculus formation will be discussed; particular emphasis is given to the inhibition mechanisms of well characterized salivary macromolecules and other exogenous moieties.

THE PRECIPITATION MEDIUM

It is not possible, at present, to characterize exactly the aqueous medium at the time precipitation of the calculus mineral takes place. However, there is enough knowledge on the compositions of saliva (5,6) and plaque fluid (7-9) - the borders of the microenvironment - to characterize these media in terms of their mineralizing potential, i.e., in terms of their DS (10,11) values with respect to mineral phases found in calculus, e.g., hydroxyapatite, $Ca_5OH(PO_4)_3$ (HA), beta-tricalcium phosphate, dicalcium phosphate dihydrate, $CaHPO_4.2H_2O$ (DCPD), and octacalcium phosphate $Ca_4H(PO_4)_3.2\ 1/2\ H_2O$ (OCP). There are still some uncertainties; salivary secretions change composition significantly (12-14) with the rate of flow. Fortunately, it appears (15) that the saturation status is similar in all stimulated salivas and distinctly different from that in unstimulated secretions. With respect to plaque fluid, PF, marked compositional differences exist between PF from active and from resting dental plaque (9, and unpublished results). The generalizations made here, however, are still valid, at least qualitatively.

Driving Forces for Precipitation

The DS value with respect to a given calcium phosphate in an aqueous medium can be expressed as the ratio of the ionic activity product (IP) for that calcium phosphate to its solubility product constant (K_s). An alternative definition, adopted in the present publication, is in terms of mean activities. Thus, as an example, the DS value with respect to HA is defined as DS $=\{((Ca^{2+})^5(OH^-)(PO_4^{3-})^3/K_s(HA)\}^{1/9}$; the denominator in the fractional power corresponds to the number of ions originating from the dissolution process or, in the case of calcium phosphates, the number of ions present in the crystalline lattice. DS values of unity, higher than unity, and smaller than unity, indicate saturation, supersaturation, and undersaturation, respectively.

From actual analytical results, it has been calculated (3) that the DS value of resting parotid saliva with respect to HA is in the order of 3.6 whereas the DS values of stimulated parotid, submandibular, and whole salivas are in the order of 21.5. Consequently, the adoption of laboratory models to study aspects of calculus formation that may be affected by salivary constituents, e.g., seeded apatitic crystal growth, requires DS values bracketed by the two extremes mentioned above. Stimulated saliva, whole and glandular, has a pH value around 7 (16) and it appears to be slightly supersaturated with respect to DCPD.

The PF appears to be supersaturated with respect to both DCPD and HA. The reported DS values with respect to DCPD are, 1.33 and 2.16 (10), 1.78 and 2.56 (11), and 2.57 (8). The reported DS values with respect to HA are, 2.99 and 4.86 (10), 11.6 and 5.20 (11), and 8.76 (8). A DS value of 16.3 has been reported (8) with respect to fluorapatite, FA. When the DS values in the cited references were not given in terms of the ratios defined here, pertinent calculations were made to express them as the mean activity ratios used in the present publication. If the enamel mineral is, as customary, considered an HA displaying a higher solubility product than that of well crystallized, pure synthetic preparations, then the PF is significantly supersaturated with respect to enamel. The DS values of the PF indicate that there is a natural tendency to form mineral within the plaque, even in an unaggregated state which, indeed, has been reported (17). The supersaturation of the PF with respect to DCPD explains the presence of this calcium phosphate in young calculus (18); the sequence DCPD---> octacalcium phosphate---> beta-tricalcium phosphate---> apatite, observed in dental calculus with increasing age (19), agrees with the increasing stability of these phosphates as indicated by the relative position of their solubility isotherms in the dilute region (20), applicable to the oral environment. The implication of the latter observation is that, after precipitation of DCPD, this phase redissolves and the next phase in the stability sequence starts being precipitated (i.e., DCPD hydrolyzes to a more basic calcium phosphate); the process repeats itself and ends with the formation of apatitic crystals having a significant content of carbonate (21).

CRYSTAL GROWTH INHIBITORS OF ORAL IMPORTANCE

Endogenous Inhibitors

The most active of the natural salivary inhibitors of crystal growth are the peptide statherine, having forty three amino-acid residues and a molecular mass of 5,388 daltons (22,23), and a group of acidic proline-rich proteins, PRP, (24) with molecular masses of 11.32 and 16 kilodaltons, that have been extensively studied in our laboratories. These proteins exhibit a very high adsorption affinity for apatitic surfaces (25,26); indeed this property is the key to their inhibitory activity. In this article, reference will be made to the salivary proteins PRP1 and PRP3 having 150 and 106 amino-acid residues, respectively. Until recently, it was thought that the primary structure of PRP3 corresponded to the first 106 residues of PRP3; however, recent studies (27) show that these proteins exhibit genetic polymorphism with respect to the residues in positions 4 and 50 which are occupied by aspartate or asparagine residues. The experiments described below using purified PRP were conducted with isomeric mixtures; however, it seems that the adsorption behavior and the inhibitory activity are quite similar for the isomeric pairs.

In typical studies on the inhibition of seeded HA crystal growth by PRP conducted in our laboratories (3,28,29), the composition of the experimental solution is 1.06 mM in Ca (from $CaCl_2$) and 0.63 mM in P (from a mixture of KH_2PO4 and K_2HPO4). The pH value is adjusted to 7.40 by the use of the appropriate proportion of the two phosphate salts. Sodium chloride, 50 mM, is used as background electrolyte. The DS value of this solution with respect to HA is 15.9 which is within the range calculated for salivary secretions. The solution volume used is usually 250 mL. The HA used as seeds in the first PRP studies reported here had a specific surface area of 19.2 m^2/g and a Ca/P molar ratio of 1.67 $\pm$.01 (30). The solution contained the inhibitor, at specified concentrations, prior to the addition of seeds. At time zero, HA seeds (25 mg) were added to the thermostated solution (37 $\pm$ 0.5° C) and the change in solution composition with time (due to crystal growth) was monitored by a continuous recording of the pH value and by periodically withdrawing small samples of suspension (5 mL) with a syringe and filtering it through a plastic-encased membrane (0.65 um pore size) attached to the syringe; the filtrate was analyzed for Ca and phosphate. The technique used, sometimes referred to as the "free drift" technique, has been described in more detail (31).

In figure 1 are shown the experimental results illustrating

the inhibition of HA crystal growth by PRP3. Similar curves were obtained when the concentration of Ca (or the pH value) in the solution was plotted against time after the addition of seeds and for the results obtained with PRP1. It is customary to fit a function of time (a polynomial) to these experimental curves in order to calculate the initial rates of precipitation, R_o, at time zero (31).

Inhibitory activity of the PRP is related (3) to the adsorption of the macromolecules onto the apatitic surface. Consequently, their inhibitory activity should be a function of adsorption coverage which can be calculated if the adsorption isotherms of the inhibitors are known. These isotherms are often described (26) by a Langmuir model of the form Q=KNC/[1+KC] in which Q is the amount adsorbed (per unit of surface area or mass), K is the affinity constant that reflects the strength of the binding bond, and N is the maximum number of adsorption sites available per unit of surface area, mols/m^2; the two adsorption parameters can be obtained through a linearalization of the model or by a non-linear least squares procedure. The fractional adsorption coverage is defined by the ratio Q/N and can be calculated from the initial concentration of inhibitor, C_o, in the experimental solution, by use of the expression Q/N = { A + (A^2-4fNC$_o$)$^{1/2}$ }/2fN in which A=C$_o$+1/K+Nf and f= ws/v; in the last expression, w is the weight of seeds, s is their specific surface area, and v is the volume of the experimental solution. That the initial rates of precipitation (crystal growth rates) are closely related to the adsorption coverage of PRP3 is shown in figure 2. In this figure, the reduction in the precipitation rate is defined as the ratio $(R_o - R_i)/R_o$, in which the subscripts "o" and "i" refer to the initial precipitation rates in the absence and presence of the inhibitor, respectively. These results are consistent with the idea that the crystal growth

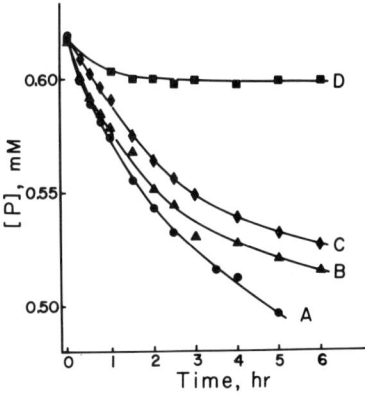

Figure 1. P concentration in solution plotted versus time. Initial PRP3 concentrations in nmols/L: A, 0; B, 47.8; C, 98.6; D, 299.

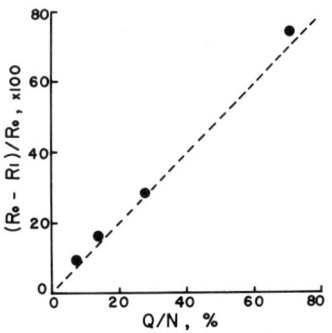

Figure 2. Reduction of initial precipitation rates as a function of PRP3 adsorption coverage. Dashed line, one-to-one correspondence.

E.C.Moreno, T.Aoba and A.Gaffar

sites are the same sites where adsorption of the protein takes place.

An important consideration relevant to the prevention of dental calculus is whether relatively bulky molecules such as PRP can block crystal growth completely, at high adsorption coverages, even when the precipitation driving forces are increased within reasonable limits. The increase in the driving force may occur, for example, by increases in the Ca and phosphate concentrations of the PF brought about by diet or by agents used in oral hygiene, e.g., dentifrices and mouth washes. A case of particular importance is the increase in the precipitation driving force by the presence of fluoride ion; in this case, the DS value increases because it should be considered with respect to a fluoridated apatite, which has a lower solubility product constant than HA. Experimentation related to this topic was carried out (28) using a solution with the nominal composition already given (actual DS value with respect to HA of 16.1) and having an initial PRP3 concentration of 3.87×10^{-5} mols/L; from the adsorption parameters, it was calculated that this concentration should cover 99.9% of the adsorption sites available for the protein adsorption on the surface of the HA seeds (25 mg and specific surface area of 16.1 m^2/g). The results obtained are shown graphically in figure 3.

Upon addition of HA seeds, there was a small change in solution composition that, probably, originates from readjustments induced by the formation of the double layer in the solution at the vicinity of the seeds in suspension. Stabilization was then attained and the composition remained constant for an observational period of eight hours. However, when a solution of NaF was added two hours after the seeds addition, enough to yield a concentration of 1 ppm in fluoride in the experimental solution (and a DS value of 37.2 with respect to fluorapatite, FA), precipitation ensued immediately as evidenced by the continued decrease of the phosphorus concentration in the solution (with concomitant decreases in Ca and pH values,

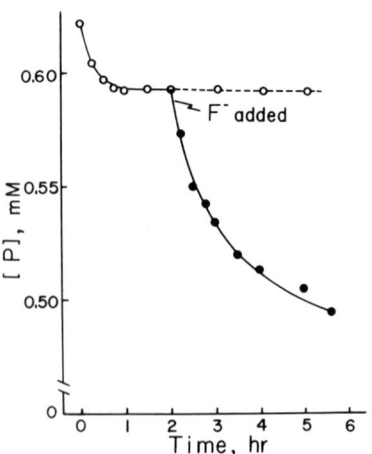

Figure 3. Initial complete inhibition of crystal growth by PRP3 is overcome by addition of fluoride (1 ppm).

not shown here). This induced crystal growth by addition of fluoride is consistent with the growth of fluoridated apatite on HA seeds and it shows that not all the crystal growth sites were blocked by the adsorbed protein. On the basis of these results, and others not reported here, it was concluded that, at maximum coverages with PRP, there is still about 15% of sites that cannot accommodate protein molecules for steric reasons but are available for crystal growth. This phenomenon is not observed with small adsorbates, as will be shown in relation to exogenous inhibitors.

The question whether the inhibitory activity of a macromolecule can be ascribed to a specific molecular segment, is often raised in the literature. In the case of the PRP and their possible role as anticalculus agents, this question has more than academic interest since it has been reported (32) that PRP incubated with saliva are degraded but inhibition of mineral formation is maintained by some of the resultant fragments. Also, it has been reported recently (33) that most of the phosphopeptides found in whole saliva derive from the amino terminus of the PRP. Furthermore, well characterized segments of PRP seem to inhibit formation of basic calcium phosphates from supersaturated solutions (4). Of particular interest is the behavior of the 30-residue amino terminus segment, obtained by a tryptic digestion (34,35) and subsequent purification. This fragment has very similar adsorption parameters to those of the parent molecule (25); also, the reduction in crystal growth rates brought about by the 30-residue segment and the parent molecule is essentially the same (29). On the basis of the foregoing information, it is tempting to conclude that the inhibitory activity of the PRP resides in the amino terminus. However, it has been shown (25,26) that the adsorption of PRP onto apatitic surfaces is entropically driven and that, most probably, dehydration of the protein and the adsorbent, as well as changes in the secondary structure of the adsorbate, account for the gain in entropy upon adsorption. Therefore, the inhibitory activity is related, as a rule, to the whole molecule and not to a portion of it. At present, it is believed that the similar inhibitory behavior of the 30-residue segment and the PRP may constitute an exceptional case in which the segment is the key to important properties such as hydrophilicity and some secondary structural features (e.g. intramolecular or intermolecular ion-pairs) displayed by the parent molecule; these properties seem to originate, to a large extent, from the presence of the two phosphoserine groups present at residues 8 and 22 (29). It is pertinent to point out that, contrary to the implication of mechanisms advanced based on simple electrostatic forces, e.g. (36), to explain the adsorption onto apatitic surfaces, adsorption is, for the PRP and other molecules, an endothermic reaction (25,26); in contrast, adsorption based exclusively on electrostatic attraction should be an exothermic process.

Exogenous Inhibitors

In recent years, special emphasis has been given to the crystal growth inhibitory activity of synthetic compounds. Inhibitory agents have found their way in the market of dentrifices and their presence in the formulation of cleansing preparations, as anticalculus agents, has been used as guarantee of the quality of the product. The effectiveness of some of these agents _in vivo_ seems to agree with their inhibitory activity of crystal growth determined _in vitro_, as reported in this meeting by Gaffar and co-workers.

As explained in the case of PRP, relatively large molecules

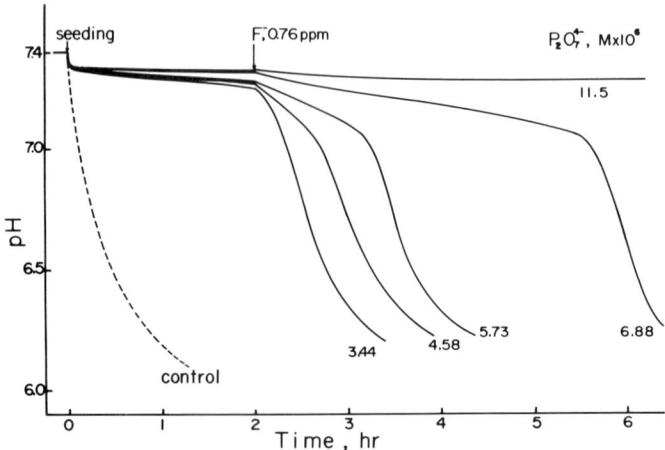

Figure 4. Effect of fluoride addition on the crystal
growth inhibitory activity of pyrophosphate present
initially at the indicated concentrations in the
supersaturated solution.

sterically block adsorption sites that remain available for
crystal growth. Consequently, such molecules can not provide
total crystal growth inhibition. With a small molecule, having a
high adsorption affinity for apatitic surfaces, it should be
possible to block crystal growth even when the precipitation
driving force is enhanced by addition of fluoride, as previously
described. One such small molecule is pyrophosphate, $P_2O_7^{4-}$, PP,
which has been reported to inhibit apatitic crystal growth
(37,38) and to adsorb onto two different HA sites of which only
one is involved in crystal growth (38).

Figure 4 illustrates the inhibitory activity of PP determined
by the use of the supersaturated solution already described
(this time, the control solution also contained fluoride at a
concentration of 0.76 ppm) to which 60 mg of HA seeds (specific
surface area 12.5 m^2/g) was added at time zero. The pH, monitored thereafter, is plotted as a function of time. Immediately
after addition of the HA seeds to the experimental solution,
crystal growth ensued as indicated by the decrease in the pH
value of the control solution, without PP. The results in figure
4 indicate that when initial PP concentration was 11.5 x 10^{-6}
mols/L (calculated coverage of 84 %), the increase in the
precipitation driving force brought about by addition of fluoride, one hour after seeding, did not translate into any crystal
growth. Such a complete inhibition contrasts sharply with the
behavior observed with the PRP and fulfills the expectations
based on the reasoning consigned previously.

An interesting feature illustrated in figure 4 is that, when
the initial concentration of PP was lower than 11.5 x 10^{-6}
mols/L, crystal growth took place following an induction period

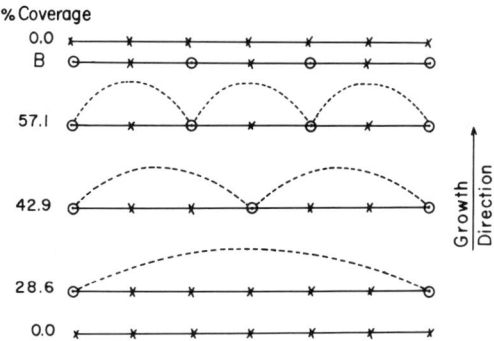

Figure 5.   Possible mechanism for induction periods.
            See text.

after the addition of fluoride; the length of this induction
period decreased with decreasing initial PP concentrations,
i.e., with decreasing adsorption coverages. One explanation (39)
for the kind of induction reported here is that during crystal
growth, the inhibitor becomes buried by advancing growing
fronts, so that a new surface is generated exempt of the inhibitor itself. This view is simplified by the two-dimensional
scheme shown in figure 5. The growing surface is represented by
a plane projection having seven adsorption sites (kinks, dislo-
cations, etc.); when an inhibitor is adsorbed in two of these
sites (a coverage of 28.6 %) the projection of the advancing
growth step describes an arc having a long radius. As coverages
increase, the curvature of the advancing steps becomes more pronounced; velocity of a spreading step is directly proportional
to the radius of the arc; for values below the "critical" curvature, the step spreads at finite velocities and after a period
of slow growth (induction), the spreading front coalesces into a
planar projection, leaving inhibitor molecules buried in a plane
behind the new spreading front represented by plane B in the
figure. In this fashion, a clean surface is regenerated and crystal growth then proceeds at rates comparable to that of the
control system, i.e., in the absence of inhibitor. For the critical and higher coverages, the spreading velocity becomes infi-
nitely slow and complete inhibition of crystal growth is attained, but this situation can happen only with small inhibitors, so that the molecules in the adsorbed state are in close
proximity.
     The foregoing mechanism for the activity of crystal growth
inhibitors of low molecular masses is qualitatively supported by
the results obtained with inhibitors having relatively high
molecular masses, e.g., the PRP; with these molecules, no induction periods have ever been observed, presumably because it is
not possible to bury the inhibitor molecules within the growing
crystal. In this context, it is of interest the results obtained
with polyvinylphosphonate, whose monomer can be represented by
$[-CH_2-CPO_3^{2-}]_n$, PVP, in which n represents the degree of polymerization. The experiments reported here were conducted with

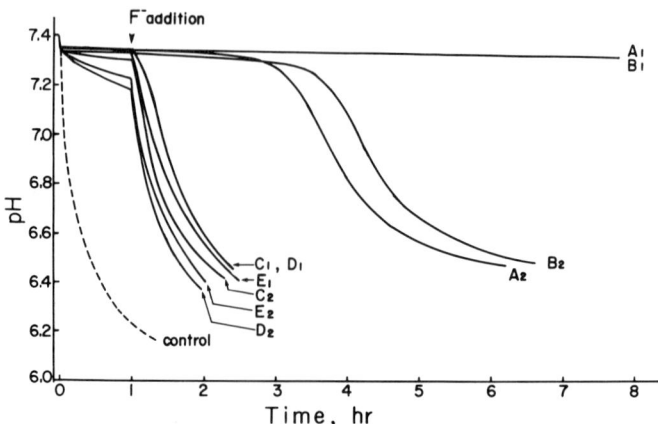

Figure 6. Effect of fluoride addition (0.76 ppm) on the
crystal growth inhibitory activity of various fractions
of polyvinyl phosphonate. Molecular weight ranges, S,
and concentrations, mg/100 mL were: Al, S<1,000 and 16;
A 2, S<1,000 and 3.2; BI 1,000<S<3,500 and 16; B2,1,000
<S<3,500 and 3.2; C1, 3,500<6-8,000 and 27.2;   C2, 3,500
<S<6-8,000 and 11.2; D1, 8,000<S<12,000 and 25.5; D2,
8,000<8<12,000 and 12.8; E1, S>12,000 and 25.5; E2, S>
12,000 and 12.8.

fractions of the polymer having the molecular masses reported on
the basis of the molecular cut off values of the dialysis membranes used for their fractionation. The supersaturated solution
had the same composition described before, but the HA seeds used
had a Ca/P molar ratio of 1.67 ± 0.02 and a specific surface
area of 10.6 m^2/g (40). At time zero, the HA seeds (30 mg/100 mL
of solution) were added and the ensuing changes in the solution
composition were monitored as previously described. Small
volumes of PVP solution (in NaCl, 0.05 mols/L and pH 7.4), at
specified concentrations, was added to the supersaturated solution before adding the HA seeds; the actual concentrations and
characterization of the PVP fractions used, are given in the
legend of figure 6.
     The results shown in this figure indicate that the two
smaller fractions (smaller than 1,000 and between 1,000 and
3,500 daltons) yielded complete inhibition of crystal growth,
even after the addition of fluoride (to yield 1 ppm F^-), when 16
mg of the polymer was present in the supersaturated solution
(curves A_1 and B_1 in figure 6); if the means of the cut off
values are taken to represent the mean molecular masses of these
two fractions, the concentrations would be 3.2 x 10^{-4} and 6.4 x
10^{-5} mols/L, respectively. For smaller concentrations of the two
low molecular mass fractions of PVP in the experimental solution, the observed effect on crystal growth reminisces that of
PP; in both cases, there is an induction period after the fluoride addition and then, precipitation proceeds at rates comparable to those observed for the control system, i.e., without in-

hibitor but having F, Ca, and phosphate in the supersaturated solution. In the case of PP, this behavior was ascribed to the burying of PP molecules by the advancing growth fronts. This explanation is difficult to visualize for larger molecules like the fractions A and B of PVP, although it could be reasoned that the burying in the latter case occurs between crystalline domains rather than a crystalline continuum. Consistent with the advanced explanation, for increasing molecular masses of PVP the observed behavior is quite comparable to that observed with the PRP, i.e., precipitation ensues immediately upon the addition of fluoride.

The foregoing results suggest that inhibitory activity is mainly determined by two factors: 1) the adsorption affinity and 2) the molecular packing on the crystalline surface. The affinity, in the case of small molecules, may be related to the hydrophilicity of the molecule (and adsorbent) since the transport of water from the adsorbed state to the bulk will increase the entropy of the system and the adsorption reaction is entropically driven. For larger molecules, changes in the secondary structure, e.g., breaking of intramolecular ion-pairs, can also contribute to gains of entropy by the system. The larger the molecule, the more probable the contribution of secondary structure to the adsorption process; however, increasing molecular size seems to bring about steric effects which impede adsorption of molecules in close proximity, thus leaving sites available for crystal growth. A third factor that appears to be important is related to an apparent specificity of inhibitors for the crystalline surface, as indicated below.

INHIBITION OF DCPD CRYSTAL GROWTH

Formation of dental calculus in the immediate vicinity of the tooth surface is expected to be determined by the composition of the PF. Since the PF appears already to be supersaturated with respect to DCPD, further increases in Ca or phosphate (or pH) concentrations will result in the precipitation of that hydrated acid calcium phosphate. The PF is also supersaturated with respect to HA but formation of DCPD is kinetically favored. It is reasonable to think, therefore, that inhibition of DCPD formation would decrease the incidence of calculus formation. One of the questions in this context is whether the agents that inhibit apatitic crystal growth are also growth inhibitors of DCPD. This question was examined in our laboratories with two inhibitors of relatively small molecular masses, PP and phytate, PY.

Crystal growth experiments were conducted using HA and DCPD seeds for comparative purposes. The solution already described (100 mL) was used when HA seeds (30 mg, specific surface area 10.6 m^2/g) were added. The initial solution concentrations of the inhibitors, in each case, was the same, namely, 1 x 10^{-6} mols/L. In figure 7A are plotted the Ca and P concentrations of the solution as a function of time after addition of the seeds. It is apparent that, whereas a significant inhibition of crystal growth was attained with PP, no significant inhibitory activity was displayed by PY. Parallel experimentation was conducted for the case of DCPD.

The supersaturated solution used in crystal growth experiments with DCPD had the following composition: Ca and P, 5.30 x 10^{-3} mols/L; pH value, 6.10; and 0.05 mols/L of NaCl as background electrolyte. This solution has DS values of 10.1 and 1.3 with respect to HA and DCPD, respectively; these DS values are sufficiently close to those reported for the plaque fluid (see above). Each inhibitor was added to the supersaturated solution at a concentration of 10^{-6} mols/L. The DCPD used as seeds was synthesized in our laboratory; it had Ca and P contents of 23.12 $\pm$ 0.05 and 18.14 $\pm$ 0.02 %, respectively, a molar Ca/P ratio of 0.99 $\pm$ 0.02, and a specific surface area (N_2) of 0.84 m^2/g. At time zero, DCPD, 200 mg, was added to the solution and the latter was monitored for changes in pH values and solution composition. In figure 7B are plotted the kinetic results obtained; the Ca curve was consistent with the P curve and with the precipitation of a phase having a Ca/P molar ratio of 1. It is clear that, in contrast with the results with HA (figure 7A), PY is a powerful inhibitor of DCPD crystal growth; in fact, its inhibitory activity surpasses that of PP in this case. These results suggest that, for the characterization of mineral accretion, it is necessary to specify not only the thermodynamic driving forces, but also the solid phase that is formed; this is particularly important in cases where the potential inhibitory activity of various compounds is to be determined.

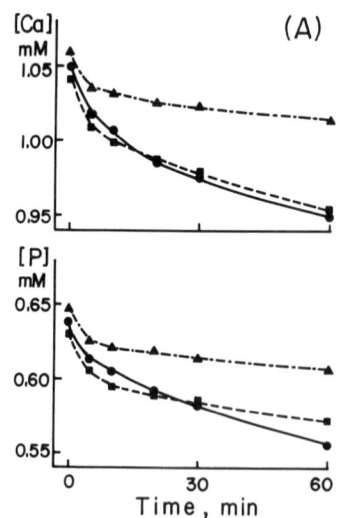 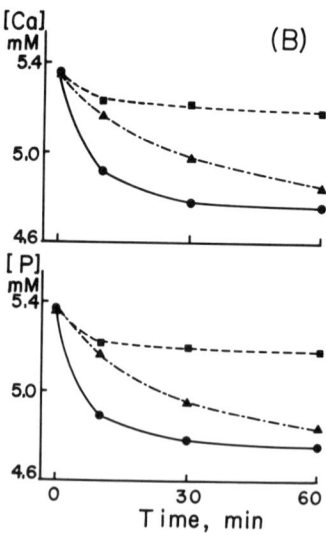

Figure 7.

A. Inhibitory effect of pyrophosphate (triangles) and phytate (squares) on crystal growth of HA.

B. Corresponding plots for inhibition of DCPD crystal growth. In both A and B initial concentraton of inhibitors was 1 micromol/L.

Acknowledgement: This work was supported by grants DE-03187 and DE-08670 from the National Institute of Dental Research and a grant from the Colgate Palmolive Co.

REFERENCES

1. Gron, P. and Hay, D.I. (1976) Archs Oral Biology, <u>21</u>, 201-205.
2. Schlesinger, D.H. and Hay, D.I. (1977) J. Biological Chemistry, <u>252</u>, 1689-1695.
3. Moreno, E.C., Varughese, K. and Hay, D.I. (1979) Calcified Tissue International, <u>28</u>, 7-16.
4. Hay, D.I., Carlson, E.R., Schluckebier, S.K., Moreno, E.C. and Schlesinger, D.H. (1987) Calcified Tissue International, <u>40</u>, 126-132.
5. Hay, D.I., Schluckebier, S.K. and Moreno, E.C. (1982) Calcified Tissue International, <u>34</u>, 531-538.
6. Dawes, C. (1984) in Cariology Today (ed. Guggenheim, B.) pp 70-74, Basel, Karger.
7. Tatevossian, A. and Gould, C.T. (1976) Archs Oral Biology <u>21</u>, 313-317 and 319-323.
8. Carey, C., Gregory, T., Rupp, W., Tatevossian, A. and Vogel, G.L. (1976) in Factors relating to demineralisation and remineralisation of the teeth, (ed. Leach, S.A.) pp 163-173. IRL Press, Oxford and Washington D.C.
9. Rankine, C.A.N., Moreno, E.C., Vogel, G.L. and Margolis, H. (1985) J. Dental Research <u>64</u>, 1275-1280.
10. Moreno, E.C. and Margolis, H. (1988) J. Dental Research, <u>67</u>, 1181-1189.
11. Margolis, H., Duckworth, J.H. and Moreno, E.C. (1988) J. Dental Research (in press).
12. Lagerlof, F. (1983) Caries Research, <u>17</u>, 403-411.
13. McCann, H.G. (1968) in Art and Science of Dental Caries Research (ed. Harris, R.S.) pp.55-73. Academic Press, New York, NY, USA.
14. Shannon, I.L., Suddick, R.P. and Dowd, F.J. (1974) in Saliva: Composition and Secretion (ed.Myers,H.M.) pp.1-96. National-Leitung AG, Basel, Switzerland.
15. Hay, D.I., Schluckebier, S.K. and Moreno, E.C. (1982) Calcified Tissue International, 34, 531-538.
16. Gron, P. (1973) Arch Oral Biology, <u>18</u>, 1365-1378.
17. Kaufman, H.W. and Kleinberg, I. (1973) Calcified Tissue Research, <u>11</u>, 97-104.
18. Schroeder, H.E. (1965) Helv Odont Acta, <u>9</u>, 73-86.
19. Schroeder, H.E. (1969) Formation and inhibition of Dental Calculus, pp 109-115, Hans Huber Publ, Berne, Switzerland.
20. Gregory, T.M., Moreno, E.C., Patel, J.M. and Brown, W.E. (1974) J. Res. Nat. Bur. Stand. - A. Physics and Chemistry, <u>78A</u>, 667-674.
21. LeGeros, R.Z. and Shannon, I.L. (1979) J. Dental Research, <u>12</u>, 2371-2377.
22. Hay, D.I. (1973) Archs Oral Biology, <u>20</u>, 1517-1529.
23. Schlesinger, D.H. and Hay, D.I. (1977) J. Biological Chemistry, <u>252</u>, 1689-1695.
24. Oppenheim, F.G., Hay, D.I. and Franzblau, C. (1971) Biochemistry, <u>10</u>, 4233-4238.
25. Moreno, E.C., Kresak, M. and Hay, D.I. (1982) J. Biological Chemistry, <u>257</u>, 2981-2989.
26. Moreno, E.C., Kresak, M. and Hay, D.I. (1984) Calcified Tissue International, <u>36</u>, 48-59.

27. Maeda, N., Kim, H.S., Azen, E.A. and Smithies, O. 91985) J. Biological Chemistry, 260, 11123-11130.
28. Margolis, H., Varughese, K. and Moreno, E.C. (1982) Calcified Tissue International, 34, S33-S40.
29. Aoba, T., Moreno, E.C. and Hay, D.I. (1984) Calcified Tissue International, 36, 651-658.
30. Moreno, E.C., Kresak, M. and Zahradnik, R.T. (1977) Caries Research, 11 (Suppl 1), 142-171.
31. Moreno, E.C., Zahradnik, R.T., Glazman, A. and Hwu, R. (1977) Calcified Tissue Research, 24, 47-57.
32. Hay, D.I. and Gron, P. (1976) in Microbial Aspects of Dental Caries, (eds. Stiles, H.M., Loesche, W.J. and O'Brien, T.C.) pp. 143-150. Information Retrieval Inc., Washington, D.C.
33. Minaguchi, K., Madapallimattam, G. and Bennick, A. (1988) Biochemical J. 250, 171-177.
34. Bennick, A., Cannon, M. and Madapallimattam, G. (1979) Biochemical J., 183, 115-126.
35. Schlesinger, D.H. and Hay, D.I. (1981) International J. Peptide Protein Research, 17, 34-41.
36. Rolla, G., Benesvoll, P. and Opermann, R. (1979) in Saliva and Dental Caries, (eds. Kleinberg, I., Ellison, S.A. and Mandel, I.D.) pp 227-241. IRL Press, New York, USA.
37. Meyer, J.L. and Nancollas, G.H. (1973) Calcified Tissue Research, 13, 295-303.
38. Moreno, E.C., Aoba, T. and Margolis, H. (1987) Compendium of Continuing Education in Dentistry, 8, S256-S266.
39. Ohara, M. and Reid, R.C. (1973) Modelling Crystal Growth Rates from Solution, Englewood Cliffs, NJ: Prentice-Hall.
40. Aoba, T. and Moreno, E.C. (1984) J. Dental Research, 63, 874-880.

R.J.Gilbert
G.S.Ingram
P.I.Riley
R.L.B.Tan-Walker

Metal ions as calculus inhibitors with particular reference to zinc

Unilever Dental Research, Port Sunlight Laboratory,
Bebington, Merseyside L63 3JW, UK

ABSTRACT

Two routes exist by which calculus formation may be inhibited. These are plaque inhibition and subsequent inhibition of mineralisation. A number of metal ions have been shown to act by both of these two mechanisms. This paper reviews the evidence for metal ion uptake to hydroxyapatite in vitro and shows that uptake of zinc is reversible. Metal ions are also shown to alter the crystallinity of hydroxyapatite in vitro. In mouth experiments show the delivery of metal ions to the oral cavity during topical application. Retention in plaque, calculus and on the toothsurface has been demonstrated. This occurs at a level sufficient to explain the clinically observed anticalculus effects of metal ions.

Recent Advances in the Study of
Dental Calculus

INTRODUCTION: CRYSTAL GROWTH AND INHIBITION

Crystal growth can be visualised to occur in three successive steps:-
● transport of lattice ions to the crystal/solution interface;
● adsorption at the interface;
● incorporation into the lattice.

Because the rate determining steps for crystal growth are frequently controlled at the crystal-solution interface, small amounts of foreign ions can have marked effects. Such substances (crystal poisons) adsorb strongly at high energy sites on the growing crystal (usually associated with crystal imperfections; edges, kinks, steps etc) and block adsorption of ions.

Depending on the type and concentration of inhibitor three effects can be observed:
● change in crystal morphology
● inhibition of crystal growth
● stabilisation of early formed phases (often amorphous)

Alternatively, crystal growth inhibition can occur when the adsorbed poison is incorporated into the growing crystal forming a solid solution which is more soluble than the pure phase. In such cases growth inhibition may result from the decrease in supersaturation. Finally, if the poison is present at a relatively high concentration then it may form water soluble complexes which again may lower the degree of supersaturation.

Crystal poisons may also stabilise supersaturated solutions against growth of nuclei by adsorbing at high energy sites. It is well established that metal ions can interfere with crystal growth causing both habit modification and total suppression of growth (1). Chromium and iron are particularly well known for this effect, causing habit modification for sodium chloride growth (2) and total inhibition of growth for ammonium dihydrogen phosphate (3), ammonium sulphate (4) and sodium chloride (2).

Crystal growth and phase transition for calcium carbonate phases are particularly relevant to geochemists and marine scientists. In such systems many metal ions (Fe^{2+}, Zn^{2+}, Ce^{3+}, Pb^{2+}, Fe^{3+}, Co^{2+}, Mn^{2+}, Be^{2+}, Ba^{2+}, Sr^{2+}, Mg^{2+}) can reduce the growth rate of calcitic phases and inhibit the aragonite - calcite phase transition (5-8).

For biologically relevant mineral phases there is substantial interest in crystal growth inhibition by metal ions particularly for calcium phosphate phases (apatite phases) and calcium oxalate phases (urinary calculi).

Much work has been directed towards Magnesium (9-10) due to its presence in biological fluids, and its competition with the chemically similar but larger calcium ion.

Many metals, but particularly Sn^{2+}, Mg^{2+}, Sr^{2+}, Cu^{2+}, Fe^{2+}, Al^{3+}, Cr^{3+} have been shown to be potent inhibitors of both seeded crystal growth from calcifying solution onto calcium phosphate or apatite seeds and to stabilise early formed amorphous phases (11-15).

In most cases the mechanism of crystal growth inhibition was postulated to be sorption of the metal ion at growth points so preventing attachment of lattice ions.

Metal ions have also been shown to be potent inhibitors of calcification of biological matrices. Thus, cadmium, manganese and cobalt inhibit cartilage mineralisation (16) while manganese inhibits collagen mineralisation (15).

Magnesium, ytterbium and zinc have been shown to be inhibitors of calcium oxalate crystal growth (17) via the formation of water soluble metal-oxalate complexes.

In addition to simple metal ions, metal-ion organic-ligand complexes can also be potent inhibitors of crystal growth with the metal ion often showing synergism with the ligand. Thus citrate has been well documented to enhance the inhibitory crystal growth activity of iron towards collagen mineralistion (15), calcium phosphate crystal growth (18) and calcium oxalate crystal growth (19). Such synergism depends on the metal, the ligand and on the metal to ligand ratio. The mechansim of synergism was not clear but was thought to involve the stabilisation of inhibitory polymeric species by ligand.

As far as dental calculus is concerned, in principle, there are two main ways by which a calculus inhibitory agent may act. These are by reduction of plaque levels (20,21), thus reducing the amount of bacterial mass available for calcification, and by inhibition or reduction of the subsequent mineralisation process (21-23). Both of these approaches have been employed successfully in the reduction of calculus formation although the most potent calculus inhibitory agents are generally mineralisation inhibitors.

METAL IONS AND DENTAL MINERAL

The calcification mechanisms by which dental plaque becomes mineralised and forms calculus have been widely described at this meeting. For the purposes of the examination of the role of metal ions in the inhibition of this process, this paper will discuss metal ions occurring naturally in calculus mineral, as substitutes in apatite lattices, as modifiers of the crystallinity of apatites, and as subsequent inhibitors of crystal growth.

Metal ions have long been known to be present as naturally occurring trace elements in a number of calcified tissues (24-28). For example Brudevold et al (24) demonstrated that zinc accumulates in human teeth with the major deposition taking place before tooth erruption and further suggested that zinc competed with calcium for positions on the apatite crystal surface. Accumulation of the divalent metal ions Ca, Mg, Mn, Sr and Zn in supra- and subgingival calculus was demonstrated by Knuuttila et al (25-27). These workers showed higher concentrations of these metal ions to be present in subgingival calculus and speculated that this was due to higher local tissue and gingival fluid concentrations of Sr and Mg than in saliva.

Mg ions are well know to be involved in plaque mineralisation due to the formation of Mg-whitlockite (29) and the similarity of the ionic radius of the zinc ion to those of Ca and Mg led to the work of Knuuttila suggesting that Zn may be some kind of nucleator in the formation of whitlockite. It was suggested that Zn-Mg whitlockites were present in human

TABLE 1.   ZINC UPTAKE BY HYDROXYAPATITE IN VITRO

Treatment	Zinc content of treated hydroxyapatite ($\mu g\ g^{-1}$)
0.132% zinc acetate	13760
0.033% zinc acetate	13620
0.013% zinc acetate	7380
Water	20

Reproduced with permission from J. Pharm. Pharmacol. (34)

subgingival calculus.

Metal ions have been implicated as potential contributors to the equilibrium between dental mineral and saliva, and as such may have a role influencing caries or calculus formation. Zinc, for example, has been shown (24) to increase resistance of hydroxyapatite to acid dissolution and Sr, Mg and Cu have all been implicated as cariostatic agents (30). Legeros (31) demonstrated that Ba, Sr, Pb, Cd were readily incorporated into synthetically precipitated apatites, while Zn, Al and Fe were capable of surpressing apatite crystal growth, also with significant incorporation. Incorporation of Sr^{2+}, Ba^{2+}, Pb^{2+}, Al^{3+} and Fe^{3+} gave an expanded a- and c- axis dimension while Zn^{2+}, Cu^{2+} and Cr^{3+} showed expanded a- and diminished c-axes.

The incorporation of metal ions also has implications for the degree of crystallinity of the material produced. For example, apatites precipitated in the presence of 0.015 mMolar Sn^{2+}, also at low concentrations, were amorphous in nature (32). Similarly, the same workers reported that reductions in apatite cyrstallinity accompanied increases in concentrations of Al^{3+}, Cr^{3+} and Zn^{2+} in the solutions from which the apatites were precipitated. Changes in concentration from zero up to 20% in the metal ion to Ca^{2+} ratio of the solution led to significant reductions in crystallinity and degree of apatite formation.

Consistent with these findings of Legeros (31) are the more recent observations of Harrap (32) who studied the effect of zinc on the transformation of amorphous calcium phosphate to hydroxyapatite. Induction time was measured by a titrimetric method (33). Zinc sulphate ($0-400\mu$ mole L^{-1}) delayed hydroxyapatite formation and this effect was enhanced synergistically in combination with sodium citrate. With each salt at $100\mu M$, induction time was delayed by 10.7 minutes compared to 3.3 minutes for the same salts individually.

ZINC UPTAKE BY HYDROXYAPATITE AND CRYSTAL GROWTH INHIBITION

In vitro experiments in our laboratory have developed the observations described above that zinc can bind to hydroxyapatite (30). Portions of 50mg of hydroxyapatite were saturated for zinc uptake by 5 minute exposure to zinc acetate solutions of above 0.033% w/w (Table 1).

TABLE 2. CONCENTRATION DEPENDENCE OF UPTAKE OF ZINC BY
HYDROXYAPATITE IN VITRO

Zinc Concn (mM)	Zinc uptake (μg g^{-1})
0.1	1412
0.25	3465
0.5	6235
1.0	7022
2.5	8649
5.0	9642
10.0	10405

TABLE 3. INHIBITION OF ZINC UPTAKE BY ADDED CALCIUM

Ca added (mM)	Zn uptake (μg/g^{-1})
0.0	449
0.5	442
1.0	384
2.5	318

The concentration dependence of this process has been
examined in more detail. The data in Table 2 shows the zinc
uptake after 2 hours incubation of zinc acetate solutions at
37°C with synthetic hydroxyapatite of surface area 18m^2/g and
Ca:P molar ratio of 1.60. Solids were separated by filtration
and washed twice with 10ml distilled water prior to analysis
of the supernatant by atomic absorption spectroscopy.
The process of zinc adsorption can be inhibited by the
presence of added calcium and is reversible. Table 3 shows
the decrease in zinc uptake with increasing calcium
concentration in the incubation and Table 4 shows that
subsequent treatment with calcium of zinc treated hydroxyapatite leads to the release of zinc from the hydroxyapatite.
In these experiments, hydroxyapatite treated with zinc was
incubated for 2 hours in the presence of different concentrations of calcium chloride. Zinc released is shown in Table
4, demonstrating the partial reversibility of the process.
This reversibility is consistent with observed clinical data

TABLE 4.   REVERSIBILITY OF ZINC UPTAKE BY HYDROXYAPATITE

Ca added (mM)	Zn removed ($\mu g/g^{-1}$)
0.0	0.13
0.5	0.34
1.0	0.51
2.5	0.80

TABLE 5.   INHIBITION OF SEEDED HYDROXYAPATITE CRYSTAL GROWTH BY ZINC

	Ca depletion (μg Ca ml^{-1} h^{-1})	% inhibition of crystal growth
a) Zinc solutions		
Water	26.0	–
0.035% zinc acetate	8.0	69%
0.131% zinc acetate	6.8	74%
0.263% zinc acetate	6.4	75%
b) Toothpaste slurries		
Non zinc citrate	28.6	–
0.5% zinc citrate paste	6.9	76%
1.0% zinc citrate paste	7.2	75%
Positive control*	7.4	74%

* Positive control was a commercially available pyrophosphate antitartar toothpaste

showing that zinc had no adverse effect on the anticaries efficacy of fluoride in a recent anticaries trial (31).

Treatment of hydroxyapatite with zinc has also been used to test the crystal growth inhibition by zinc of seeded hydroxyapatite crystal growth (36,37). Hydroxyapatite (50mg, Ca:P(M) = 1.61) of surface area $26m^2$ g^{-1} was pretreated for five minutes with test solution (20ml) or toothpaste extract (20ml of 1:4 suspension of paste in water), washed and resuspended in 50ml amounts of calcifying solution (Ca 1.5mM, PO_4 4.5mM, pH 7.4). The depletion of calcium from the solution was then monitored over 1hr as a measure of crystal growth. Table 5 below shows the crystal growth inhibition achieved by test solutions containing zinc and by slurries of dentifrice containing zinc citrate.

These findings show that a five minute exposure of hydroxy-apatite to zinc solutions of concentrations representing dilutions of 0.5% zinc citrate toothpaste (0.131% zinc acetate is equivalent to 10g 0.5% zinc citrate paste plus 30g water) can reduce substantially the seeded crystal growth of the hydroxy-apatite in a calcifying solution. Similar findings are demonstrated for treatment with slurries of toothpastes containing 0.5% and 1.0% zinc citrate.

ORAL PHARMACOLOGY OF METAL IONS

The efficacy of any preventive or drug treatment is dependent on that agent being available to reach the site of action in sufficient concentration for sufficient time to exert its biological effect. This section will review some of available data concerning the delivery and biological effect of metal ions in the oral cavity with particular reference to agents which may have an anticalculus action.

Amongst the earliest records of the intra oral effects of preparations containing metal ions is the work of Hanke (38) who demonstrated antibacterial properties of mercury, copper, silver, zinc and nickel salts. Hanke also observed the importance of oral retention time in conferring efficacy by increasing the time of contact between the metal ion preparation and the prospective site of action.

Oral retention has also been demonstrated for a number of metal ions with observable antiplaque efficacy. Many of these ions are also species demonstrated to have crystal growth inhibitory properties in vitro. Retention of 37% of the tin in 10mls of a 22.2mM SnF_2 solution was shown by Bonesvol and Rolla (39). This level of retention corresponded to 0.82μmoles of tin and these authors concluded that tin was retained in high amounts after use of mouthrinses but that salivary tin levels fall more rapidly than those of the antiplaque agent chlorhexidine. Uptake of tin into dental plaque (40) and inhibition of bacterial acid production (39, 41) have been demonstrated and proposed to account for the antiplaque effect of this metal ion. Subsequently, Attramadal and Svatun (42) reported rapid uptake of tin by Streptococcus mutans in vitro in a process apparently independent of cell metabolism. The level of uptake of ca 1mMole Sn/g cells dry weight is several orders of magnitude greater than the levels of Sn (0.015mM) shown to reduce apatite crystallinity (31). Furthermore these workers (42) reported that most of the bound tin was present extra cellularly, making it more likely that such material would be available to interfere with crystal growth in vivo.

Oral retention of copper and zinc following mouthrinsing has also been demonstrated. Afseth et al. (43) showed 31% retention for copper and 15% for zinc after use of a mouthrinse containing 0.1M copper-sulphate and a 2.5M zinc acetate mouthrinse. In a comparable study, Harrap et al (44) reported that 12% of the zinc from a 30mM zinc phenolsulphate mouthrinse was retained in the mouth after expectoration of the mouthrinse.

While oral retention data demonstrate residence in the

mouth, it is important for clinical efficacy that adequate concentrations of active agents are achieved at the sites of action, i.e. for anticalculus agents in plaque, calcifying plaque or calculus.

Both fluoride and none-fluoride salts of copper, tin and silver have been shown to reach plaque and to inhibit plaque acid production in vivo following a sucrose challenge (45,46). Silver and copper were more active at inhibiting plaque acid production than tin but this difference could not be related only to oral retention levels. Of the two potential routes by which metal ions may have anticalculus activity it has been suggested by these and other authors (47) that an important determinant of efficacy against plaque metabolism is metal ion affinity for bacterial enzyme sulphydryl groups. Several enzymes involved in the bacterial glycolytic pathway contain such -SH groups. In terms of effectiveness for the inhibition of mineralisation, once again the plaque levels reported (45, 46) for copper and zinc of above 200µg metal ion/g plaque are well in excess of those capable of inhibiting apatite crystal growth in in vitro systems (ca 10µg g^{-1} reference 31).

Zinc has also been reported to be retained in the mouth (in plaque and other oral sites) following use of mouthrinses (44, 48) and toothpastes (34,49,50) containing levels of zinc which have been reported to have therapeutic antiplaque and anti-calculus efficacy. Harrap, Saxton and Best, for example, carried out two studies (44, 48) on mouthrinses containing various zinc salts at different concentrations. These authors found that solutions containing zinc as the citrate or phenol-sulphonate salts of concentration 17-30mM gave about 30% reduction in plaque growth. Following use of these preparations saliva zinc levels were significantly increased for up to four hours after dosage. Initial post dosage saliva zinc levels were of about 200 times the normal and while this had dropped to ca 12 times normal an hour later, only after 4 hours did the zinc level in saliva reach the upper end of the range of values found for background zinc levels. No differences in zinc retention were observed between use of citrate, sulphate or phenolsulphonate zinc salts. In these experiments, zinc levels were also found to be raised in plaque for up to six hours after treatment. Concentrations of just under 500µg zinc per gram dry weight of plaque were present 6 hours after dosage. No estimates are available of the total quantity of zinc found in plaque after use of oral dosage vehicles apparently because no one has attempted to sample all the plaque from the mouth and express that as a proportion of the dosage quantity.

There are a number of dentifrices widely available containing zinc salts as an adjunct to plaque and tartar control. Some studies are also available which demonstrate oral delivery of zinc from these preparations (34, 49, 50).

Reports of the oral retention of zinc from toothpastes containing 0.5% (w/w) zinc citrate have varied from 24% (50) to 38% (34) of the zinc dosed during brushing with 1g of the test dentifrice. In each of these cases salivary zinc levels were elevated for at least 2 hours after brushing. A pharmacokinetic analysis of saliva zinc levels (50) reported

TABLE 6. ZINC IN CALCULUS AFTER 3 MONTH USAGE OF DENTIFRICES
CONTAINING ZINC CITRATE

Treatment Dentifrice	µgZn/Mg calculus	Molar ratio Ca:Zn in calculus sample
Control (n=53)	0.22 ± 0.03	4200:1
0.5% zinc citrate (n=19)	0.51 ± 0.50	2535:1
1.0% zinc citrate (n=30)	1.06 ± 0.17	524:1

an elimination constant of 0.0147 min^{-1} with a biological
half- life (t $1_{/2}$) of 47 minutes. This figure is indicative
of the widespread binding and retention of zinc in the oral
cavity. Similar saliva zinc levels were reported in a further
study in which 2g of the 0.5% zinc citrate toothpaste was
used. In this study Saxton et al (49) found measurable plaque
zinc levels above background for four hours after use of this
test toothpaste. This observation has been confirmed by
Gilbert and Ingram (34). Zinc was found to be present in
elevated concentration in both plaque fluid and plaque residue
(49). This finding could be important as an indication of
the presence of soluble zinc in plaque fluid rather than
insoluble zinc bound by bacteria or material of the plaque
matrix.

The level of zinc measured in plaque after use of an
dentifrice containing 0.5% zinc citrate has been also compared
with that required to inhibit seeded hydroxyapatite crystal
growth in in vitro tests, Ingram and Carter (36) found that
pretreatment of hydroxyapatite with aqueous slurries (4:1,
v/v) of the test toothpaste substantially inhibited crystal
growth. They also found that pretreatment with solutions
containing 0.035% zinc acetate had the same effect. This
level of zinc acetate gave zinc uptake to hydroxyapatite of
13,620µg zinc per gram of hydroxyapatite (34).

Dry plaque contains ca 8µg calcium per mg (51). This is
equivalent to 20µg of hydroxy apatite. The levels of zinc in
wet plaque was reported to be 0.82µg mg^{-1} (34) corresponding
to approximately 4µg zinc mg^{-1} dry weight. Thus, on the basis
of the amount of calcium phosphate present, an uptake of 200mg
zinc g^{-1} hydroxyapatite is estimated. Although some zinc may
be adsorbed elsewhere in plaque, this figure of 200mg zinc g^{-1}
represents a substantial excess over those levels shown to be
acquired by hydroxyapatite and to inhibit its crystal growth
in vitro.

Similarly, levels of zinc have recently been measured in
calculus of human volunteers participating in clinical trials
of zinc containing dentifrices (52). Table 6 shows the zinc
levels recovered from calculus from individuals who had used
dentifrices containing 0.5% or 1.0% zinc citrate for 3 months.
Once again, levels of zinc recovered are well in excess of
those shown to inhibit cyrstal growth in vitro.

These observations are consistent with the in vitro observations described earlier that zinc can replace calcium in the calcium phosphate crystal matrix.

In conclusion, this paper has reviewed the evidence for metal ions as crystal growth inhibitors. Particular reference to zinc has shown that this metal ion is bound reversibly by hydroxyapatite and inhibits seeded hydroxyapatite crystal growth. Furthermore, in common with many metal ions, zinc is retained in the mouth after topical application in appropriate quantities and at appropriate sites to explain the observed anticalculus effects of this ion.

REFERENCES

1. Kirk-Othmer Encyclopedia of Chemical Technology Vol 7 Third Edition, John Wiley and Sons Inc.
2. Cooke, E.G., Kirst Tech., 1, 119 (1966).
3. Davey, R.J. and Mullin, J.W., J. Cryst. Growth, 26, 45 (1974).
4. Larson, M.A. and Mullin J.W., J. Cryst. Growth, 20, 183 (1973).
5. Meyer, H.J., J. Cryst. Growth, 66, 639 (1984).
6. Kitano, Y., Kanamori, N. and Yoshioka, S., Geochem. J., 10, 175 (1976).
7. McLester, M.E., Martin, D.F. and Taft, W.H., J. Inorg. Nucl. Chem., 32, 391 (1970).
8. Kitano, Y., Kanoriori, N. and Tokuyama, A., Am. Zoologist, 9, 681 (1969).
9. Nancollas, G.H., Tomazic, B. and Tomson, M., Croatica Chemica Actu., 48, 431 (1976).
10. Amjad, Z., Koutsoukos, P.G. and Nancollas G.H., J. Colloid and Interface Sci., 101, 250 (1984).
11. Meyer, J.L. and Nancollas, G.H., J. Dent. Res. Sept/Oct. 1443 (1972.
12. Bachra, B.N. and Fischer, H.R.A., Calcif Tiss. Res. 3, 348 (1969).
13. Bachra, B.N. and van Harskamp, G.A., Calcif. Tiss. Res. 4, 359 (1970).
14. Meyer, J.L. and Angino, E.E., Invest. Urol. 14(5) 347 (1977).
15. Thomas, W.C., Proc. Soc. Exptl. Bio. and med. 170, 321 (1982).
16. Bird, E.D. and Thomas, W.C., Proc. Soc. ExpH. Bro. and med., 112, 640 (1963).
17. Leskovar, P., Kratzer, M. and Baustadter, R., Therapiewache, 30, 4291 (1980).
18. Meyer, J.L. and Thomas, W.C., J. Urol. 128(6), 1372 (1982).
19. Meyer, J.L. and Thomas, W.C., J. Urol. 128 1376 (1982).
20. Volpe, A.R. et al. (1965) J. Periodont. 41, 463-467.
21. Stephen, K.W. et al. (1987) Caries Res. 21, 380-384.
22. Schiff, T.G. (1986) Clin. Prev. Dent. 8, 8-10.
23. Zacherl, W.A. (1985) J.A.D.A. 110, 737-738.
24. Brudevold, F. et al. (1963) Arch. oral Biol. 8, 135-144.
25. Knuuttila, M. et al. (1979) Scand. J. Dent. Res. 87, 192-196.
26. Knuuttila, M. et al. (1981) Scand. J. Dent. Res. 89, 412-416.

27. Knuuttila, M. et al. (1980) Scand. J. Dent. Res. <u>88</u>, 513-516.
28. Knychalska-Darwan, Z. et al. (1985) Folia Histochemica Cytobiologica <u>23</u>, 21-26.
29. Gron, P. et al. (1967) Arch. Oral Biol. <u>17</u>, 829-837.
30. Legeros, R.Z. et al. (1976) Calc. Tiss. Res. <u>22</u>, 362-367.
31. Legeros, R.Z. et al. (1980) Proc. 2nd. Int. cong. Phosphorus Compounds.
32. Harrap, G.J. (1988) J. Dent. Res. <u>67</u>, (special issue) 320.
33. Briner, W. and Francis, M. (1973) Calcif. Tiss. Res. <u>11</u>, 10.
34. Gilbert, R.J. and Ingram, G.S. (1988) J. Pharm. Pharmacol. <u>40</u>, 399-402.
35. Stephen, K.W. et al. (1988) Comm. Dent. Oral. tpidemiol (in press).
36. Ingram, G.S. and Carter, P. (1987) J. Dent. Res. <u>66</u>, (special issue) 198.
37. Ingram, G.S. et al. (1988) J. Dent. Res. <u>67</u>, (special issue) 402.
38. Hanke, M.T. (1940) J.A.D.A. <u>27</u>, 1379-1393.
39. Bonesvoll, P. and Rolla, G. (1978) Caries Res. <u>12</u>, 112.
40. Svatun, B. and Attramadal, A. (1978) Acta. Odontol. Scand. <u>36</u>, 211-218.
41. Lilienthal, B. (1956) Aust. Dent. J. <u>1</u>, 165-173.
42. Attramadal, A. and Svatun, B. (1980) Acta. Odontol. Scand. <u>38</u>, 349-354.
43. Afseth, J. et al. (1983) Scand. J. Dent. Res. <u>91</u>, 42-45.
44. Harrap, G.J. et al. (1984) Arch. Oral. Biol. <u>29</u>, 87-91.
45. Opperman, R.V. and Johansen, J.R. (1980) Scand. J. Dent. Res. <u>88</u>, 476-480.
46. Afseth, J. (1983) Scand. J. Dent. Res. <u>91</u>, 169-174.
47. Oppermen, R.V. and Rolla, G. (1981) in Dental Plaque and Surface Interactions (ed. leach, S.A.) pp 225-234. IRL Press, Oxford, UK.
48. Harrap, G.J. et al. (1983) J. Periodont. Res. <u>18</u> 634-642.
49. Saxton, C.A. et al. (1986) J. Clin. Periodont. <u>13</u>, 301-306.
50. Gilbert, R.J. (1987) J. Pharm. Pharmacol. <u>39</u>, 480-483.
51. Jenkins, G.N. (1978) in Physiology and Biochemistry of the Mouth. Chapter 10 p.372 Blackwell, Oxford, UK.
52. Ingram, G.S. et al. (1988) Caries Research (in press) ORCA abstract.

A.Gaffar
T.Aoba
J.Afflitto
A.Esposito
E.C.Moreno

Structure-activity relationship between *in vitro* inhibition of HA crystal growth and *in vivo* anti-calculus effects

Colgate-Palmolive Research Center, 909 River Road,
Piscataway, NJ 08854-5596, USA and
Forsyth Dental Center, 140 Fenway, Boston,
MA 02115, USA

ABSTRACT

We have examined low mol. wt. and macromolecules as the
inhibitors for their effect on hydroxyapatite (HA) formation,
crystal growth in vitro and have correlated the parameters
with the in vivo anticalculus effects in rats or beagle dogs.
A series of carboxyphosphonates were used to ascertain this
relationship. These are phosphonoformate (PFA), phosphono-
acetate (PAA), phosphonopropane (PPT) and phosphonobutane
(PBTA) carboxylic acids. Additionally, the homopolymer of
polyvinylphosphonic acid (PVPA) and polyallylphosphonoacetate
(PAPA) were also investigated for assessing the relationship
between the influence of molecular weight on the in vivo
anticalculus effect. HA formation was studied in a supersaturated Ca and PO_4 solution (4×10^{-3}M) at constant pH 7.4
in a nitrogen atmosphere. The seeded crystal growth was
studied using Ca, 1.06mM and P, 0.63mM in 0.05M NaCl at pH
7.4. The growth was induced by adding well characterized HA
seeds. PFA, PAA, PPT and PBTA inhibited HA formation at 13,
28, 6.2 and 3.0×10^{-5}M respectively. The topical application
of the sodium salts of the compounds in rats or beagle dogs
kept on a calculogenic diet gave a significant reduction in
calculus vs. placebo; for PFA (37%), PPA (58%), PPT (30%) and
PBTA 80 percent,respectively. The macromolecules PVPA and
PAPA inhibited the crystal growth of HA at 0.02 and 0.7
micromoles respectively. The topical applications of 1%
solutions of the polymers gave 16 and 30 percent reductions
for PVPA and PAPA respectively. The strength of binding, K,
of the molecules and the number of available sites (N)
for binding to HA were also estimated. PFA, PPT or PVPA did
not increase caries in vivo nor did they adversely affect the
anticaries efficacy of fluoride ion in vivo.

Recent Advances in the Study of
Dental Calculus

INTRODUCTION
    Dental calculus is a mineralized deposit that forms within
and around dental plaque. It can occur either subgingivally
or supragingivally in the vicinity of the plaque although the
presence of the latter is not a prerequisite. Calculus can
form in the absence of plaque (1) and a number of anti-plaque
agents are effective in inhibiting plaque in vivo without
affecting calculus formation (2). Conversely, most of the
effective inhibitors of calculus in humans do not inhibit
plaque formation (3).
    Dental calculus consists of both organic and inorganic com-
ponents. The organic portion is a combination of epithelial
cells, leukocytes, microorganisms and polysaccharides. The
inorganic part is primarily calcium phosphate salts which
include carbonated hydroxyapatite (HA), dicalcium phosphate
dihydrate (DCPD), and octacalcium phosphate (OCP). A general
method for removing calculus is by mechanical scraping by a
dentist or hygienist. This approach, albeit painstaking, is
widely used. Another approach would be to develop agents
which selectively inhibit the formation of calculus minerals,
i.e., the hard crystalline dental deposits.
    A large number of compounds have been investigated to inhi-
bit calculus formation in animals and man. Among the effective agents are pyrophosphate (PPi), pyrophosphate analogs,
diphosphonates and polyphosphonates (4). These agents inhibit
calculus formation by inhibiting crystal formation of calcium
phosphate salts such as HA. In this report, we describe the
crystal growth inhibiting effects of several carboxyphosphonates in vitro and in vivo. The effects of these compounds in
the presence of fluoride was also assessed. Carboxyphosphonates are analogs of phosphocitric acid, a known regulator of
the mineralization process in vivo (5).

MATERIAL AND METHODS
    Trisodium phosphonoformate (PFA) and sodium phosphonoacetate
(PAA) were obtained from Richmond Organics (Virginia, USA).
Sodium phosphonobutane tricarboxylic acid (PBTA) was supplied
by the Mobay Chemical Company (Pittsburgh, PA, USA). Sodium
phosphonopropane tricarboxylic acid, (PPT), was obtained from
Henkel (Dusseldorf, Germany). OTB was obtained from the
Monsanto Company (St. Louis, MO). Sodium polyvinylphosphonic
acid (PVPA) and polyallylphosphonacetate (PAPA) was
synthesized in our laboratories. The structures of the
compounds are in Table 1.

Synthesis of PAPA
    For the synthesis of PAPA, a solution of 50 grams of allyl
alcohol and 24 ml of phosphonoacetic acid in 500 ml of acetonitrile was refluxed overnight in an anhydrous atmosphere
under a soxhlet extractor containing 3°A molecular sieves.
The solution was cooled, and AIBN (2,2',-azobis-2-methyl
proprionitrile) was added. The resulting solution was stirred
under anhydrous nitrogen at 74°C. Successive 3g. portions of
AIBN were added at 24 and 48 hrs. After 72 hrs., the mixture
was cooled and the supernatant was decanted from the crude

Table 1. Structures of Inhibitors

Phosphonoformic Acid (PFA)	$H_2O_3P-COOH$
Phosphonoacetic Acid (PAA)	$H_2O_3P-CH_2-COOH$
1 Phosphonopropane 1,2,3 Tricarboxylic Acid (PPT)	$H_2O_3P-CH(COOH)CH(COOH)CH_2COOH$
2 Phosphonobutane 1,2,4 Tricarboxylic Acid (PBTA)	$CH_2(COOH)C(PO_3H_2)(COOH)-CH_2-CH_2-COOH$

Polyallylphosphonoacetate (PAPA)

$$[CH_2-CH]_n$$
$$CH_2-O-\overset{O}{\overset{\|}{C}}-CH_2-\overset{O}{\underset{OH}{\overset{\|}{P}}}-OH$$

Polyvinylphosphonic Acid (PVPA)

$$\left[CH_2-CH\right]_n$$
$$HO - P = O$$
$$OH$$

Malonic Acid (MA)	$COOH-CH_2-COOH$

Pyrophosphate (PPI)

$$HO - \overset{O}{\underset{OH}{\overset{\|}{P}}} - O - \overset{O}{\underset{OH}{\overset{\|}{P}}} - OH$$

3-Oxa-2,2,4, Tricarboxybutan-1-ol (OTB)	$HO-CH_2C(COOH)_2-O-CH(COOH)CH_3$

polyallylphosphonoacetate. The polymer was washed with
several small portions of acetonitrile and then placed under
0.1 Torr vacuum for 12 hr. to remove solvent traces. The
42.06 g. of crude polymers thus obtained was dissolved in
about 800 ml of water, adjusted to 8.0 with sodium bicarbo-
nate, and dialyzed using a 3500 Dalton cut off membrane. The
retenate solution was evaporated to about a 100 ml volume and
then freeze-dried to obtain 7.5 g of pure PAPA as the sodium
salt. Proton NMR indicated the product to be at least 98%
pure with less than 1% monomers (allylphosphonoacetate) and no
detectable allyl alcohol.

Synthesis of PVPA
    A solution of 2,2-azobis (2-methyl propionitrile) (2.04g,
0.0124M) in vinylphosphonyl dichloride (60.0g) was stirred and
heated at 70°C under nitrogen for 18 hrs. The resulting solid
was ground and triturated twice with ether to remove unreac-
tive monomer and initiator. The polyvinylphosphonyl chloride
was hydrolyzed adding the powder portionwise to 40ml of water
with cooling. The hydrochloric acid and most of the water was
removed by heating at 40°C under vacuum. The pH of the residue was adjusted to 7.4 by adding 5.0M sodium hydroxide. The
polymer was dialyzed overnight using a dialysis membrane
having a 3500 Dalton cut off. The solid was recovered by lyo-
pholization. The ^{31}P NMR solution spectrum in D_2O showed a
major peak (70%) at 2239 Hz (Varian FT 80A, carrier frequency
32.198 MHz) characteristic of the polymers. The molecular
weight of the polymer by the light scattering method was 12000.

Analysis of Carboxyphosphonates

An ion chromatography method (Dionex, Sunnyvale, CA) was used to analyze PFA, PPT, PBTA, PAA and PAPA in the solutions. For the separation of different compounds, the AS7 column was used with 70 mM HNO_3 as an eluant. The post column reagent was 1 g. $Fe(NO3)_3 \cdot 9 H_2O$ in 1.0 liter of 20% $HClO_4$. The efflu-ent was analyzed by UV detector at 330nm. The typical reten-tion times were 3.42 min., 3.71 min., 4.18 min. and 3.37 min. for PFA, PAA, PPT and PBTA respectively.

Inhibition of Spontaneous Hydroxyapatite Formation

The in vitro formation of hydroxyapatite (HA) was studied via a pH-stat procedure described previously (6). Stock solu-tions of 0.1 M $CaCl_2$ and 0.1 M NaH_2PO_4 were freshly prepared in carbonate-free deionized distilled water. The reaction ves-sel contained 23 ml of deionized, distilled, and CO_2-free water. One ml of the phosphate stock solution was then added; the phosphate solution contained inhibitors at various concen-trations. Then, one ml of the calcium stock solution was added and the reaction time was counted from this point on. Thus, the calcium and phosphorus initial concentrations were both 4 mM. These concentrations fall in the range reported for plaque fluid (7). The reaction was run at pH 7.4 on a pH-stat apparatus (Radiometer, Copenhagen) in a nitrogen atmos-phere and the base consumption (0.1 N NaOH) was recorded auto-matically. All reactions were run at 25+0.05°. The time of formation, as referred to in this paper, represents the time elapsed between the addition of the calcium stock solution and the time when the solid formed exhibited an apatitic nature as indicated by X-ray and chemical analyses.

Samples of reacting suspension were taken by means of a sy-ringe when turbidity was first apparent and thereafter as de-scribed before (6). The suspension was passed through a fil-tering device attached to a syringe having a cellulose acetate membrane (Millipore) with pores of 0.65 um in diameter. The solid retained in the filter was dried and analyzed for cal-cium and phosphorus. Also, powder x-ray diffraction patterns were obtained from these early precipitates.

Seeded Crystal Growth of HA

The HA used as seeds was synthesized by a slow precipitation procedure (8); its crystallographic properties (x-ray and opti-cal) correspond to those of a pure, well-crystallized HA with a specific surface area of 16.1 m^2/g.

The experimental approach was similar to that previously de-scribed in the studies of crystal growth of HA (8). Briefly, a solution (250 ml) supersaturated with respect to HA was pre-pared from stock solutions of KH_2PO_4, K_2HPO_4 and $CaCl_2$. These solutions, although metastable, did not produce any spontane-ous precipitation during a period of several days; neverthe-less, fresh solutions were prepared for every experiment con-ducted. The solution was thermostated at 37° + 0.02 C.

The initial composition of the solution was $\overline{C}a$, 1.06 x 10^{-3}M; P, 0.63 x 10^{-3}M; NaCl, 0.05 M; and pH 7.4. The DS (degree of supersaturation) values are defined as the ratio of the ionic activity product of HA to its solubility product constant. The experimental solution had a DS value of 7.45 x 10^{-10}. The solubility product constant used for HA was 7.36 x 10^{-60} (9). The DS value of the solutions with respect to HA was within the range of supersaturation that has been estimated for unstimulated and stimulated human salivas (10). All solutions were in 50 mM NaCl to simulate the ionic strength of saliva. Upon addition of apatite seeds (25 mg) to the experimental solution, crystal growth occurred and the progress of this process could be followed by monitoring the decrease in Ca, P and pH in the solution.

The initial precipitation rate was calculated by first fitting a polynomial in time t of the form $Y_t = \sum_{i=0}^{n} A_i t^i$ in which Y_t represents the concentration of phosphorus or calcium at time t and the four coefficients A_i are those determined by the least squares procedure used for the fitting. In all cases the correlation co-efficients were higher than 0.98. The polynomial thus obtained was differentiated with respect to time, and the initial precipitation rate, R_o, was taken as the numerical value of the derivative for t = 0. Since adsorption isotherms on HA were not available for PAA, PFA, PBTA, PPT, and PAPA, the attempt was made to estimate the parameters using the obtained kinetic data, according to the following equation $C_o/\theta = (1/K) + 1/(1-\theta) + NS/V$ where,

$\theta = (R_c - R_i)/R_c$ where θ was expressed in terms of fractional reduction of the initial precipitation rates, where R_c and R_i are the initial precipitation rates in the absence and the presence of inhibitor, respectively.

S is the total surface area of the seeds used (0.742 M^2), and V=250 ml. The least square analysis of C_o/θ versus $1/(1-\theta)$ showed good linear relationship (r=0.940). The parameters K and N were obtained from the slope and intercept of the linear regression.

Effects of Inhibitors on Calculus Formation in Beagle Dogs:
Pure-bred beagle dogs were obtained from LRE (Kalamazoo, Michigan, USA). The animals were 4-5 years old, in good general health and were routinely fed Purina dog chow. Prior to the start of the study, they were given a complete dental prophylaxis which included scaling of teeth followed by polishing with a rubber cup and pumice. The beagles were kept on a calculogenic diet prepared by mixing 3.5 kg of Purina chow, 7.6 liters of water and 1 kg of special (bone fragment and bone dust-free) horse meat; the mixture was allowed to stand for 30 minutes to soften.

The experimental solutions PFA, PAA, and PBTA were applied topically as the sodium salt at pH 7.0. The calculus formation was evaluated as described in (11). The treatment phase lasted for 3-4 months. The data were analyzed by an analysis of variance followed by the Student Neuman Keul's test.

Topical Effects of PBTA on Calculus Formation in Rats
This study was conducted on litters (3 animals each) of Osborne-Mendel rats. The rats were fed a calculogenic diet consisting of the diet 580F described by Regolati et al., (12)

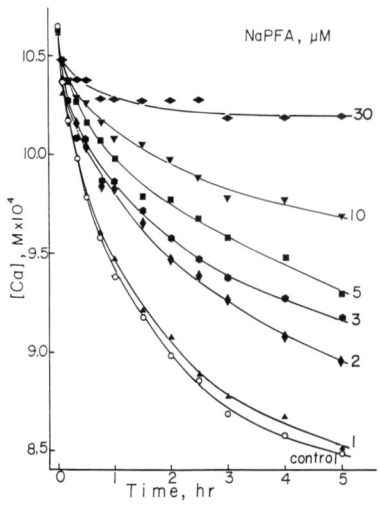

Figure 1:
Concentration of Ca in solution
plotted versus time after
addition of HA seeds to the
supersaturated PFA. The number of
each curve represents the
concentration of inhibitor in
micromoles/L.

supplemented with 0.2% P as Na_2HPO_4. On day 20, the animals
were randomly distributed among three treatment groups (two
rats per stainless steel, screen bottom cage) and began receiving calculogenic diet and tap water ad libitum. On days
21 and 22, the rats were inoculated twice daily with a heavy
suspension of Actinomyces viscosus OMZ-105-Nyl. From day 23
on, 100 ul of a 1% solution of inhibitors were applied with
disposable syringes twice daily; water was used as a control.
The treatments were delivered at 9:00 and 15:00 hours during
the 30-day experimental period. The animals were weighed at
the beginning and end of the study. Calculus formation was
assessed by the procedure described by Regolati et al. (12).
The data from the studies were analyzed by an analysis of variance followed by the Student Newman Keul's test.

Topical Effect of Inhibitors on Caries Formation in Rats
     The topical anticaries effects of PFA, PPT and PVPA were
evaluated in rats. The experiments were conducted in Osborne-
Mendel rats. The animals received tap water ad libitum and
cariogenic diet (2000a) containing 56% sucrose.
     On day 13, the animals were transferred with their dams
to stainless steel, screen-bottom cages and were fed finely
powdered Nafag stock diet and tap water ad libitum until day
20. The rats were then distributed at random among the
treatments. They continued to receive the cariogenic diet and
tap water ad libitum. On days 21 and 22, they were inoculated
twice daily with heavy suspensions of S. mutans OMZ-176 and A.
viscosus Ny-1. From day 23 to the end of the study, 100 ul of
the test dentifrices were applied twice daily using disposable
syringes. The incidence of fissure and smooth-surface caries
were assessed according to the method described previously
(13). The extent of caries was evaluated as follows. Under a
dissection microscope the extent of smooth-surface lesions on
the buccal surfaces of the first and second mandibular molars

Table 2. Effects OF PFA, PAA, PPT, PBTA, PVPA, and PAPA on HA Formation

Agent	Conc M	Time for HA Formation	Delay in HA Formation
-----	0	18 min.	-----
PFA	1.3×10^{-4}	37 min.	19 min.
PAA	2.8×10^{-4}	96 min.	77 min.
PPT	6.2×10^{-5}	62 min.	44 min.
PBTA	3×10^{-5}	78 min.	60 min.
PVPA	1.5×10^{-4}	30 min.	12 min.
PAPA	2.7×10^{-5}	48.0 min.	30 min.
Malonic Acid	2×10^{-4}	18.0 min.	none
PPI	1×10^{-4}	66.0 min.	48.0 min.

Solution composition: Ca, 4×10^{-3}M; P, 4×10^{-3}M; pH 7.4

was recorded using linear area units (E) on scales of 0 to 6
for the first molar and 0 to 4 for the second molar. After-
ward, serial mesiodistal sections were prepared from the mandi-
bles with an internal-rim containing machine. The sections
were stained, serially mounted on slides, and randomized before the evaluation on a scale of 0 to 3:in grade 1, only the
enamel was stained, but the stained areas did not involve the
dentinoenamel junction; in grade 2, only the enamel was
stained, but the stained areas included the dentinoenamel junction and not the dentin proper; in grade 3, the enamel and dentin were both stained. The means were computed by summing the
scores divided by the total units at risk. The data were analyzed by an analysis of variance followed by the Student-Neuman-
Keul's test.

RESULTS
        The results in Table 2 show the effects of the inhibitors
on HA formation in a highly supersaturated system. The most
effective inhibitor was PBTA which gave inhibition at $3 \times$
10^{-5}M, followed by PAPA showing inhibition at 2.7×10^{-5}M.
PFA and PAA exhibited an inhibitory effect at least one order
of magnitude less than PBTA, PPT or PAPA. PVPA, on the other
hand, showed an inhibitory effect at a high concentration, 1.5
x 10^{-4}. Since the concentration of calcium in the system was
4 mM, it was not clear whether the effect of PVPA was due to
true inhibition of formation of HA or due to the reduction in
the activity of calcium ion in solution which could reduce the
driving force of precipitation. Malonic acid, even at $2 \times$
10^{-4}M, failed to show an inhibitory effect.
        The effects of inhibitors on seeded crystal growth kine-
tics of HA are shown in Figures 1 to 5. A complete inhibition
of the crystal growth occurred at 30, 1.3, 2.0, 0.2, 0.7 and
0.02 micro moles respectively for PFA, PPA, PPT, PBTA, PAPA
and PVPA (Table 3). The corresponding value for pyrophosphate
(PPi) is 5.7 micromoles which has been reported previously
(14).

PA A

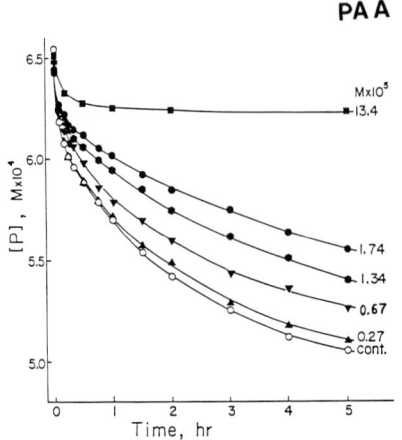

Figure 2:
Effects of PAA on seeded crystal growth of HA. The concentration of P is plotted versus time.

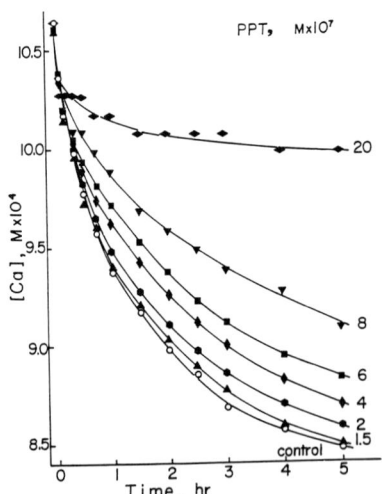

Figure 3:
Concentration of Ca plotted versus time after addition of HA seeds to the supersaturated solution containing PPT. Numbers of each curve represent the concentration of inhibitor (M x 10E-7).

The effects of topically applied PFA, PAA, OTB and MA on calculus formation in rats is shown in Table 4. At concentrations of 1% each of the compound at pH 7.0, only PFA and PAA gave significant reduction in calculus formation when compared to the control. The reductions were 37 and 58% respectively for PFA and PAA. OTP and MA were found to be ineffective. The influence of topically applied PPT, PBTA, PAPA and PVPA on the calculus formation in rats is presented in Table 5. PPT, PBTA and PAPA gave comparable but significant reductions in calculus formation in rats (28-33%) when applied at a concentration of 1%. PVPA, on the other hand, gave an 16% reduction in calculus formation. Since the experiment was done separately on each compound, it was not possible to ascertain

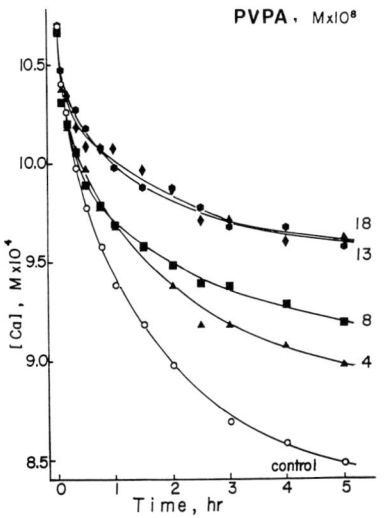

Figure 4:
Effects of PVPA on seeded crystal
growth kinetics of HA. The number
of each curve represents the
concentration of inhibitor
(M X 10E-8).

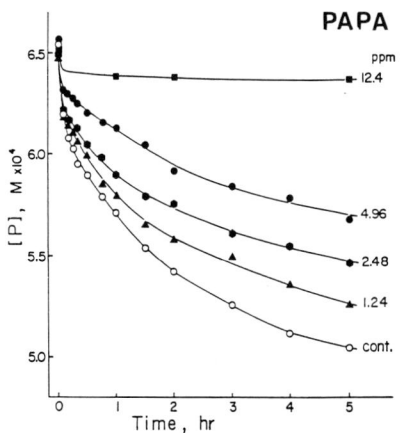

Figure 5:
The inhibition of PAPA of HA
crystal growth kinetics as
measured by change in phosphate
in solution versus the time after
addition of HA seeds.

whether or not differences exist among the low molecular
weight, carboxyphosphonates and the macromolecular inhibitors,
i.e., PAPA or PVPA. The results in Table 6 describes the
effects of PFA, PAA and PBTA on calculus formation in dogs.
PBTA provided the largest reduction in calculus formation.

Although it may not be possible to precisely state conclusions regarding the relationship between the molecular size of
the inhibitors, their adsorption parameters and in vivo efficacy, some of the adsorption parameters estimated from the
kinetic runs are given in Table 7. The values for PVPA and
PPi were directly estimated from the adsorption isotherm on
HA. The parameters for PAPA were not estimated since the precise molecular weight of the compound was unavailable at the

Table 3. Inhibition of Crystal Growth of HA by Carboxyphosphonates and the Polymers

Agent	Complete Inhibition of Growth, for 5 hr (10^{-6}M)
PFA	30
PPA	1.3
PPT	2.0
PBTA	0.2
PAPA	0.7
PVPA	0.02
PPI	5.7

Ca; 1.06×10^{-3}M; P; 0.63×10^{-3}M; HA Seeds 25 mg; pH 7.4.

Table 4. Effects of PFA, PPA, and MA Applied Topically on Calculus Formation in Rats

Treatment	N	Mean Calculus/Rat + SD	% Reduction	Sig.
Water Control,	12	7.0 + 3.4	-----	-----
1% Phosphono-formate, pH 7.0	12	4.4 + 2.9	-37.0	P ≤ 0.05
1% Phosphono-acetate, pH 7.0	12	2.9 + 2.7	-58.5	P ≤ 0.05
Malonic Acid, pH 7.0	12	6.8 + 5.3	-----	N.S.
OTB	12	7.9 + 3.6	-----	N.S.

20 Calculus Units at Risk/Rat

Table 5. Effects of Topically Applied PPT, PBTA, PAPA and PVPA on Calculus Formation in Rats

Treatment	Conc	N	Mean Calculus/Rat + SD	% Reduction	Sig.
PPT	1%	15 15	Water = 12.6 + 4.6 PPT = 8.8 + 4.7	29.6	P=0.05
PBTA	1%	12 12	Water = 11.8 + 3.1 PBTA = 7.9 + 2.5	33.0	P<0.05
PAPA	1%	15 15	Water = 17.0 + 5.7 PAPA = 12.3 + 8.3	27.6	P=0.05
PVPA	1%	15 15	Water = 30.8 + 4.0 1% PVPA = 26.0 + 5.3	16.0	P=0.05

Table 6. Effects of Topically Applied PFA, PAA and PBTA for 3 Months on Calculus Formation in Beagle Dogs

Treatment	Conc	N	Mean Calculus/Tooth + SD	% Reduction	Sig.
PFA	1%	10 10	Placebo 3.7 + 2.0 PFA 2.5 + 2.4	33.4	P = 0.05
PAA	1%	10 10	Placebo 3.0 + 2.8 PAA 1.5 + 2.0	50.0	P = 0.05
PBTA	1%	2 2	Water 4.0 + 0.0 PBTA 0.63 + 0.25	84.0	P ≤ 0.01

Table 7. Comparison of Adsorption Parameters of Inhibitors and In Vivo
Anti-Calculus Effects

| Inhibitor | Mol-Wt | Adsorption Parameters | | Functional Groups | Residual Carboxyl | In Vivo Cal. Reduct |
		N,umol/m^2	K, ml/umol			
PFA	126	0.56	170	1	1	33-37
PAA	140	0.49	59	1	1	50-58
PPT	256	0.29	4900	1	3	33
PBTA	270	0.23	2200	1	3	37 to 84
PVPA	12,000	0.038	1950	60	--	18
PPI	174	0.43	K_1=3.13 K_2=1.64X10^{-4}	2	0	58

time of the experiments. In general, there is a tendency in N
value to decrease with the increase in molecular size. For example, PVPA has an N value of 0.038 umol/m^2 which is an order
of magnitude lower than the low molecular weight inhibitors.
The lower N seems to indicate lower efficacy against calculus
formation in vivo. The values of K vary tremendously among
the molecules having approximately similar size. The molecules having multiple charged groups had high K values. However, a high K value did not necessarily result in higher efficacy against calculus in vivo.

Since the carboxyphosphonates inhibit HA formation, they
may interfere with the remineralization process in vivo.
Their effect in the presence of fluoride ion was, therefore,
assessed in the seeded crystal growth kinetic experiments.
Figure 6, illustrates the HA precipitation (as indicated by pH
change) with PFA co-existing with fluoride. In this figure,
the dashed line represents the reaction course with uncoated
seeds but having the inhibitor in the supersaturated solution
at the concentrations indicated in the graph. The solid lines
indicate the reaction course using precoated seeds (in the 10
ml solution). In these systems, using PFA or PPT as inhibitors, very high concentrations were necessary to provide high
coverage of seed crystals. Addition of 1 ppm fluoride to the
solution induced immediately the precipitation reaction rates
in both "coated and uncoated" systems except with a high concentration of PFA or PPT (Figure 6 & 7). On the other hand,
fluoride induced rapid precipitation with PVPA (Fig. 8),
following an induction period. It is apparent that the rate
of precipitation induced by the addition of fluoride depends
on the concentration of the inhibitor in the solution pre-
existing with fluoride.

Table 8 summarizes the topical effects of fluoride or fluoride plus PFA or PPT on dental caries in rats. Compared to
the water control, fissure caries were reduced (P_F≤0.05) in
the 300 PPM F/NaF, 300 PPM F + 1% PFA or 300 PPM F$^-$ plus 1%
PPT. There were no significant differences among the three
groups. The solutions containing PFA or PPT, each without
NaF, caused no significant reduction of fissure lesions compared to the control.

Table 8. Influence of PFA and PPT on Dental Caries in the Presence of Fluoride

Treatment	N	Dentinal Fissure Lesions/Rat	Smooth Surface Lesions/Rat	Weight Gain
Control	12	2.7	10.8	78
300 ppm F/NaF	12	1.0	4.0	80
1% PPT	12	2.5	10.2	77
1% PPT + 300 F	12	1.3	5.3	79
1% PFA	12	1.9	12.4	76
1% PFA + 300 PPM F	12	1.1	3.9	77
S$\bar{x}$ Standard Error of Means		0.49	1.34	3.6
S$\bar{d}$ Standard Error of Differences		0.70	1.89	5.2
P$_F$ <		0.01	0.001	n.s.
LSD 0.05		1.38	3.76	10.2

Compared with the water control, smooth surface caries were reduced (P$_F$<0.001) in the group treated with 300 PPM F/NaF. The solutions containing PPT, and PFA without fluoride neither significantly reduced nor increased smooth surface caries. When each of the solutions was applied with 300 PPM F$^-$, both PPT and PFA significantly decreased (P$_F$<0.01) caries incidence in comparison with the control. However, versus the 300 PPM F treatment alone, these caries reductions were not significant.

Compared to the water control, the fluoride content was increased (P$_F$<0.001) in 300 PPM F treatment group. Neither PFA or PPT impaired fluoride uptake in vivo (Table 9).

Table 10 shows the effects of PVPA on caries and fluoride uptake. Compared to the water control, PVPA did not increase smooth or fissure surface caries. Additionally, PVPA did not impair fluoride uptake in vivo (Table 9).

DISCUSSION

The HA formation test used in our investigations is relevant for assessing the anticalculus effects of agents since it simulates the concentration of calcium and phosphates that are found in plaque fluid (15). In this test HA formation occurs through a precursor phase, a phase more acidic than apatite, and the inhibitors tend to stabilize the phase. PBTA and PAPA were most effective in vitro under the conditions of the test exhibiting inhibition at 2.7 and 3.0 x 10^{-5}M respectively. The least effective was PVPA, PAA and PFA. The inhibitory effect of the active agents is due to the surface effect of the molecules on the precursor phases rather than the direct chelation of calcium in the solution used for conducting the test. The ratio of calcium ions to the inhibitors is too high (100:1) to account for the chelation effect being the primary mechanism. The most likely explanation is that the inhibitors affect the conversion of the precursor phases to HA. Indeed

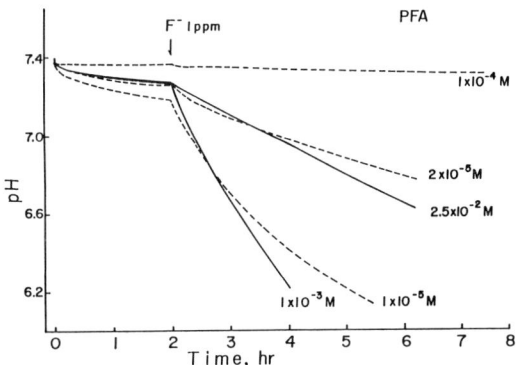

Figure 6:
Effects of fluoride addition on HA growth in systems containing
PFA in the supersaturated solution (broken lines) using uncoated
seeds and in systems with PFA in solutions but using precoated
seeds (full lines). The number of each curve represents
concentration of inhibitor in solution.

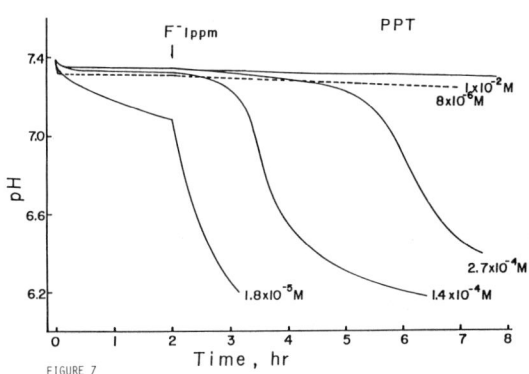

Figure 7:
HA crystal growth in system containing PPT and fluoride. A system
containing PPT in supersaturated system (broken lines) using
uncoated seeds and in systems with PPT in solutions (full lines)
but using coated seeds.

Blumenthal et al. (16) found that HA formed in the presence of
1-hydroxy-ethyledine,1,1,diphosphonate was only partly crystalline, a finding consistent with the above reasoning.
    In a seeded crystal growth kinectics of HA, the inhibi-
tory activity of various carboxyphosphonates was also assessed
(Table 3). The test simulates the conditions occurring in
saliva since the driving force of the reaction (degree of
supersaturation with respect to HA) was comparable to that

Table 9. Influence of PFA and PPT on Fluoride Uptake In Vivo

Treatments	PPM F in Layers		
	I	II	I + II
Control	46	19	32
300 PPM F⁻/NaF	263	75	170
1% PPT	46	22	34
1% PPT + 300 PPM F⁻/NaF	188	69	129
1% PFA	60	28	45
1% PFA + 300 PPM F⁻/NaF	207	71	135
Sx̄ Error of Means	18.8	6.8	12.2
LSD 0.05	52.4	19.0	34.2
$P_p <$	0.001	0.001	0.001

existing in saliva and the experimental conditions were those
previously shown to result in HA crystal growth without a precursor phase. Under these conditions, the most effective inhibitor was PVPA (2.0×10^{-8}M) followed by PBTA (2×10^{-7}M) and
PAPA (7×10^{-7}M). The reason for this differential effect
could be due to the differences in the degree of supersaturation of three orders of magnitude between the crystal growth
inhibition experiments versus the HA formation test (10^{10} vs.
10^{13}).

It is interesting to consider the structure requirement that
can be derived by comparing various carboxyphosphonates. With
other inhibitors of crystal growth such as the oligomer of poly-
sulfoacrylic acid and phosphocitric acid, it has been suggested that the size of the molecules, molecular substitu-
tions, the number of active groups, the geometrical arrangement of the groups and stability of the molecules are
important for the effect (17; 18).

From the study of the above compounds, it is quite clear that
a minimum requirement is a presence of a phosphonate and a carboxylate on the same carbon since neither malonic acid (two
carboxylate on the same carbon) nor 3-oxa-2,2,4 tricarboxybu-
tanol-1 (OTB) showed inhibitory effects on the formation of HA
in vitro. It could be that the presence of phosphate or phosphonate group in the molecule imparts the "bis" function-
ality, i.e., under typical adsorption conditions to HA or its
precursor phase two protons may become dissociated, yielding a
tetrahedrally shaped anion. This spatial arrangement may
impart a very important ability. When any one oxygen anion
encounters a cation at the surfaces, it is only necessary for
the electrons to "flow" to the most favorable distribution and
not for a group to physically turn to bring two negative
charges, with the same group, to bear upon any one cation,
allowing the formation of more than one bond to an anion.
This property is not exhibited by the carboxylate group. The
above reasoning presupposes that the inhibitors adsorbed to
the surface cation, i.e. calcium. An increase in the number
of anionic groups per se in the molecules seems to increase
the inhibitory effects of carboxyphosphonates against

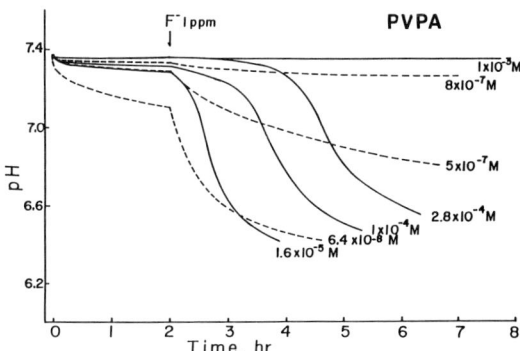

Figure 8:
Influence of PVPA on HA crystal growth in the presence of
fluoride. Broken lines indicate PVPA in supersaturated solution
using uncoated seeds and full lines indicate systems with PVPA in
solution but using precoated seeds.

hydroxyapatite formation (compared PAA vs. PBTA, PPT and
PAPA). The increase in the number of phosphonate groups,
however, in vicinal phosphonates, does not enhance the
inhibitory effect on HA formation (compared in Table 2 PAPA
vs. PVPA). This indicates that in addition to total anionic
groups in the molecules, the proximity of the other anion and
stereochemistry is an important factor in the inhibition.
Contrary to HA formation, the number of anionic groups in the
molecules seems to enhance the crystal growth inhibitory effect
of HA in vitro (compared PVPA vs. PPA, PPT, PBTA in Table 3).
The reason for this differential effect is not clear; it could
be related to the differential adsorption parameters to HA
crystals vs. precursor phases of HA.
   It is important to consider the type and location of calcium
phosphate salts when evaluating the anticalculus effect in
vivo. Dental calculus in humans is composed mainly of four
types of calcium phosphate salts: dicalcium phosphate dihydrate, OCP, magnesium containing tricalcium phosphate, and carbonate containing hydroxyapatite (19). Calculus deposits in
beagles contain a mixture of calcite and HA. In rats,
calculus is primarily HA (20). In spite of these differences,
the results obtained with both animal models correlate well
with the results obtained from human clinical studies. With
respcect of carboxyphosphonate, it has been shown that PPT
added at 1% in a dentifrice significantly reduced calculus
formation in humans by 33 percent when compared to the placebo
(21). In our test in rats, the same concentration applied
topically gave a 33 percent reduction. It is interesting to
note that PPT, PBTA AND PAPA (Table 5) applied topically gave
approximately the same degree of inhibition in vivo although
there was one order of magnitude difference in the
concentration required for inhibiting HA between PAA, PPT and
PBTA. This is consistent with previous studies with geminal

Table 10. Effects of PVPA and PVPA Plus Fluoride on Caries in Rats

Treatment	N	Advanced Dentinal* Fissure Caries	Smooth Surface** Caries	PPM F in Layers I	II	I & II
Deionized Water	12	9.1	18.8	29	7	18
200 PPM F/NaF	12	4.7	8.1	254	53	153
1% PVPA	12	8.9	18.2	65	11	39
1% PVPA + 200 PPM F	12	6.3	11.0	254	52	157
$\overline{Sx}$ Standard Error of Means		0.68	1.26	18.7	5.4	11.7
$\overline{Sd}$ Error of Differences Between Means		0.96	1.78	26.5	7.7	16.5
$P_F<$		0.001	0.001	0.001	0.001	0.001
LSD		1.92	3.55	53.0	15.3	33.0

*12 Fissures at Risk; **20 Units at Risk

and polyvicinal diphosphonates in animals. For example, it
was found in the previous studies in rats that, although
trisodium methanecyclohexylhydroxydiphosphonate (MCHDP)
inhibited HA formation at 2.0×10^{-3}M vs. Na_2EHDP at
2×10^{-4}M, these compounds gave similar anticalculus results
in rats when incorporated in the diet of the animals (22). It
should be emphasized that this was also true when a comparison
was made between Na_2EHDP and ethylenediamine(N,N,N N)tetrame-
thylene phosphonate ^{2}(Editempa) at an equimolar concentration.
Although Editempa gave an inhibition of HA formation at a
concentration of two orders of magnitude higher than Na_3 EHDP
(2×10^{-6}M vs. 2×10^{-4}M), when these agents were applied
topically at 0.05M in rats, they gave equivalent anticalculus
effects (23). On the other hand, there was a tendency towards
a higher reduction in calculus formation in beagles with the
more effective inhibitor of HA formation in vitro (Table 6).
It could be that the calculus formed in the beagles is more
heterogeneous (mixed phases) versus the calculus formed in rats
(more homogenous). It should also be emphasized that in
addition to the intrinsic potency of the inhibitor in vitro,
other pharmacokinetic factors such as intra-oral retention,
binding to and release from the tooth surfaces, and salivary
fluid flow rate may play an important part in projecting in
vitro data (obtained in a closed system) to the in vivo
efficacy (24).
When the application of a potent inhibitor of HA formation is
considered as an anticalculus agent, it is important to understand not only their potential as an inhibitor of the normal
mineralization process but also their inhibitory activity in
the presence of co-existing fluoride. Indeed, previous
studies have indicated that some anticalculus agents can interfere with the post-eruptive maturation of enamel as well as
with the mineralization process in rodents (25). It was also
found that this adverse effect could be overcome by the addition of fluoride. Considering these findings, it was of

considerable interest to assess the effects of selected carboxyphosphonates on caries in rats. The data clearly indicate
that PFA, PPT or PVPA did not increase caries in developing
enamel nor did they interfere with the anticaries effect of
fluoride in vivo. These results are consistent with recent
findings in humans. For example, in a two year caries
clinical study in children a combination of pyrophosphate/
fluoride was as effective in providing caries protection as
fluoride by itself (26). In addition, a combination of
another potent HA inhibitor, AHP (azocycloheptanone 1,1,diphos-
phonic acid) plus fluoride was more effective in caries prevention than fluoride by itself (27).
A concomitant reduction of caries and calculus when two
agents coexist in the same system can be explained by the fact
that calculus formation occurs on the teeth (above pellicle),
while the demineralization occurs in the subsurface region of
the enamel (under pellicle). The presence of pellicles on the
tooth dictates a selective transport of fluoride and the inhibitors. Also, the studies on the natural inhibitor of crystal
growth in saliva indicated that its crystal growth inhibitory
effect can be overcome by the addition of fluoride. This
effect was neither due to the displacement of the adsorbed
inhibitor by fluoride nor to the activation of secondary
growth sites. Rather the effect was explained on the basis of
an increased precipitation driving force and the incomplete
blockage of crystal growth sites on the basis of steric
reasons (28). Our in vitro and in vivo data are consistent
with the above rationale.
In summary, the results of present investigations in vitro
and in vivo indicate that low molecular weight carboxyphosphonates and macromolecules were effective inhibitors of HA formation in vitro and were effective as anticalculus agents in
vivo. In addition, these compounds did not damage developing
enamel in vivo nor did they interfere with the anticaries
effect of fluoride in vivo.

ACKNOWLEDGEMENTS
We would like to thank Mrs. Maryanne Clark for her skillful
typing of this manuscript and Dr. Orum Stringer for the
synthesis of PAPA.

REFERENCES
1.   Glas, J. E. and Krasse, B. (1962) Biophysical Studies on
      Dental Calculus from Germ Free and Conventional Rats.
      Acta Ondont. Scand. 20, 127-134.
2.   Grossman, E., Reiter, G. and Sturzenberger, P.O. (1986)
      Six Month Study of Effects of Chlorhexidine Mouthwash on
      Gingivitis in Adults. J. Periodont. Res. (Supp. 16) 21,
      33-39.
3.   Suomi, J. D., Horowitz, H. S., Barbano, M. A., Spolsky,
      W. W., and Heifetz, S. B. (1974) A Clinical Trial on
      Calculus Inhibiting Dentifrice. J. Periodontol. 45,
      139-145.
4.   Francis, M. D. (1962) Inhibition of Calcium Hydroxyapa-
      tite Crystal Growth by Polyphosphonates and Polyphos-

phates. Calcif. Tissue Res. 3, 151-162.

5.   Lehninger, A. L. (1980) Synthesis and Characterization of Phosphocitrate, A Potent Inhibitor of Hydroxyapatite Crystal Growth. Biochemistry 19, 1983-88.

6.   Gaffar, A. and Moreno, E. C. (1985) Evaluation of 2-phosphobutane 1,2,4 Tricarboxylate as a Crystal Growth Inhibitor In Vitro and In Vivo. J. Dent. Res. 64, 6-10.

7.   Tatevossian, A., and Gould, C. T. (1976) Methods of Sampling and Analyzing the Aqueous Phase of Dental Plaque. Arch. Oral Biol. 21, 313-317.

8.   Moreno, E. C., Zahradnik, R. T., Glazman, A., and Hwu, R. (1977) Precipitation of Hydroxyapatite from Dilute Solution upon Seeding. Calcif. Tissue Res. 24, 47-57.

9.   Moreno, E. C. and Varughese, K., (1981) Crystal Growth of Calcium Apatite from Dilute Solutions. J. Crystal Growth 53, 20-30.

10. Hay, D. I., Schluckebier, S. K., and Moreno, E. C. (1982) Equilibrium Dialysis and Ultrafiltration Studies of Calcium and Phosphate Binding of Human Salivary Proteins. Implications of Salivary Supersaturations with Respect to Calcium Phosphate. Calcif. Tissue Res. 34, 531-538.

11. Gaffar, A., Moreno. E. C., Muhlemann, H. R. and Niles, H. P. (1983) Effects of Editempa on Dental Calculus and Caries Formation In Vivo. Calcif. Tissue Int. 53, 362-365.

12. Regolati, B., Schmid, R., and Muhlemann, H. R. (1970) The Effects of Diphosphonate, Pyrophosphate and Sodium Fluoride on Drinking Habits of Osborne-Mendel Rats. Helv. Odont. Acta 14, 34-36.

13. Gaffar, A., Schmid, R., Afflitto, J. and Coleman, E. (1987) Effects of Pyrophosphate/Copolymer/NaF on Dental Calculus and Caries Formation In Vivo. Comp. Cont. Ed. Dent. Suppl. 8, 5251-5255.

14. Moreno, E. C., Aoba, T., and Margolis, H. C. (1987) Pyrophosphate Adsorption onto Hydroxyapatite and Its Inhibition of Crystal Growth. Comp. Cont. Ed. Dent. Suppl. 8, S256-S266.

15. Duckworth, J. H., Margolis, H. C. and Moreno, E. C. (1988) Composition of Pooled Resting Plaque Fluid from Caries Free and Caries Susceptible Individuals. J. Dent. Res. 67, 1660.

16. Blumenthal, N. C., Betts, F, and Posner, A. S. (1977) Stabilization of Amorphous Calcium Phosphate by Mg and ATP. Calcif. Tissue Res. 23, 245-250.

17. Gaffar, A., Niles, H. P. and Davis, C. B. (1981) Evaluation of an Oligomer or an Oligomer plus Cetyl Pyridinium Chloride Against Plaque, Stain, Calculus and Gingivitis. J. Dent. Res. 60, 1432-1438.

18. Williams, G. and Sallis, J. D. (1979) Structure Activity Relationship of Inhibitors of Hydroxyapatite Formation. Biochem. J. 184, 181-184.

19. Schroeder, H. and Bambauer, H. V. (1966) Stages of Calcium Phosphate Crystallization During Calculus Formation. Arch. Oral Biol. 11, 1-8.

20. Legeros, R. I. and Shannon, I. L. (1979) The Crystalline Components of Human Calculi: Humans vs. Dogs. J. Dent.

Res. <u>50</u>, 2371-2377.
21. Cassesse, G., Celeste, G., DeNotaris, V., Gargiulo, and Silvano, G. (1980): Valcitazione di una Pasta Dentifricia Contenente Come Inhibitore Dello Sviluppo del Tartaro un Derivato Fosfonic, 11 PPT, Medical Praxis. Riv. Internaz Med. Terapia e Farmocol. Clin. <u>1</u>, 221-228.
22. Briner, W. W. and Francis, M. D. (1973) <u>In Vitro</u> and <u>In Vivo</u> Evaluation of Anticalculus Agents. Calcif. Tissue Res. <u>11</u>, 10-22.
23. Gaffar, A. (1979) Unpublished data.
24. Bonesvoll, P., and Gjermo, P. A. (1978) Comparison Between Chlorhexidine and Some Quaternary Ammonium Compounds with Regard to Retention, Salivary Concentration and Plaque Inhibiting Effect in the Human Mouth After Mouthrinses. Arch. Oral Biol. <u>23</u>, 289-294.
25. Briner, W. W., Francis, M. D. and Widder, J. S. (1971) Factors Affecting the Rate of Post Eruptive Maturation of Dental Enamel. Calcif. Tissue. Res. <u>1</u>, 249-256.
26. Triol, C. W., Ripa, L. W., Leske, G. S., and Volpe, A. R. (1988) Clinical Study of the Anticaries Efficacy of Three Fluoride Dentifrices Containing Anticalculus Ingredients: One and Two Year Results. J. Clin. Dent. (in press)
27. Koch, G., Karlsson, R., Bergman-Arnodottir, I., Bjarnason, S., Finnbogason, S., and Hoskuldsson, O. (1988) A Three Year Controlled Clinical Trial on Caries Preventing Effect of Fluoride Dentifrices With or Without Anticalculus Agents (diphosphonates). In this symposium.
28. Margolis H., Varughese K., Moreno, E. C. (1982) Effect of Fluoride on Crystal Growth of Calcium Apatites in the Presence of a Salivary Inhibitor. Calcif. Tissue Int. <u>34</u>, 33-40.

D.J.White[1]
W.D.Bowman[1]
G.H.Nancollas[2]

Physical-chemical aspects of dental calculus formation and inhibition: *in vitro* and *in vivo* studies

[1]The Procter & Gamble Company, 11511 Reed Hardman Highway, Cincinnati, Ohio 45241, USA and [2]SUNY at Buffalo, Buffalo, NY, USA

Abstract

Supragingival dental calculus is the end product of mineralization of dental plaque. This mineralization process involves the development of localized supersaturation, nucleation, crystal growth and the phase transformation of mineral phases with time. The interruption and regulation of calculus can involve the reactivity of anti-calculus species at various stages of plaque calcification. Factors influencing the reactivity of these antitartar compounds can include adsorption on mineral surfaces, chelation properties in plaque, diffusion into and out of plaque, and combined reactivity of the antitartar agent with fluoride. Accurate predictive studies on the effects of antitartar compounds must take into account these complexities. This study reports our findings on the effects of antitartar agents like pyrophosphate and zinc on mineral deposition and calculus development, both *in vitro* and *in vivo*. *In vitro* studies using the constant composition technique show differences in inhibitor efficacy for the mineralization of various tartar mineral phases including apatite, DCPD and OCP. Overall, the inhibition of DCPD, a faster growing mineral phase, requires greater ambient solution concentrations of inhibitor than that required for HAP or OCP. An exception to this was pyrophosphate which similarly inhibited HAP and DCPD mineralization at concentrations of 1-2 μMol/L. Extended mineralization studies show overgrowth, 2° nucleation and inhibitor degradation to be factors limiting TC agent efficacy. *In vivo* studies using the rat calculus model demonstrate differences in inhibitor effects on plaque mediated mineralization with test results similar to those obtained in *in vitro* screens of TC dentifrices using the CC technique.

INTRODUCTION

Dental calculus results from plaque mineralization. Minerals identified in human calculus include amorphous calcium phosphate (hereafter ACP), dicalcium phosphate dihydrate (DCPD), octacalcium phosphate (OCP), magnesium substituted tricalcium phosphate (whitlockite- WH) and hydroxyapatite (HAP) (1-5). In general, more acidic precursor phases are identified in early plaque calcification (primarily DCPD, ACP & OCP) and more basic phases (HAP, WH) are found in mature tartar (2,4,6-9). While many approaches have been considered for the control of calculus development (antiplaque agents, solubilizing agents-chelators, antiadhesion agents etc. (2,10-14)), the most successful approach to calculus prevention to date has centered upon the utilization of crystal growth inhibitors, like the diphosphonate molecules studied in the mid 1970's (10,15-19). Tartar control (TC) agents used in commercial dentifrices or mouthrinses today also include mineral growth inhibitors such as soluble pyrophosphate salts (20-23), zinc chloride (24) and zinc citrate (25-26). While the general relationship between the chemical inhibitory action of these agents and clinical anticalculus activity is empirically well established (16) there still remains lack of knowledge about the precise chemistry which may differentiate efficacy for various agents. In particular, considerable uncertainty exists about the relationship between solution chemical effects of TC agents and clinical measures of calculus, which involve the prevalence or "area coverage" measurements of calculus on the teeth.

Traditional _in vitro_ methods for the evaluation of tartar control actives have included measures of agent effects on spontaneous and seeded pH stat mineralization of HAP (16,27-30). In these studies, the mineralizing solutions are initially supersaturated with respect to all four calcium phosphate phases. Under such conditions the crystallization process can involve the nucleation, growth, and re-dissolution of kinetically favored precursor phases (28, 31-32). Furthermore, the large changes in solution supersaturation during precipitation can change the overall kinetic processes dominating mineralization reactions (33). Lastly, the decreasing driving force during mineralization prohibits the accurate assessment of overgrowth, inhibitor breakdown, secondary nucleation or other effects which might contribute to benefits or limitations to inhibitor actives.

The constant composition (CC) technique provides important advantages for the assessment of mineralization kinetics of calcium phosphates, overcoming many limitations of conventional methods (34-37). By virtue of its control over solution thermodynamic conditions, the CC method enables the controlled study of inhibitor influences on the mineralization of calculus precursor mineral phases such as OCP and DCPD which may play critical roles in plaque petrification. In addition, the ability to grow minerals for extended periods provides the opportunity to quantitatively examine processes contributing

to limitations of inhibitor agents, including overgrowth and
secondary nucleation effects. As a result, the CC method
would appear to be an ideal adjunct to traditional in vitro as
well as in vivo methods toward the assessment of tartar control active effects on calculus processes.

The purpose of this study was to 1) examine and compare
the reaction profiles of commercially used inhibitor agents,
including zinc, zinc citrate, pyrophosphate and EHDP for
various calculus mineral phases, including DCPD, OCP and HAP
using the CC technique (34-38), 2) preliminarily examine limitations of agents, such as inhibitor overgrowth, secondary
nucleation, or inhibitor breakdown which may influence clinical effects of these agents in practice, and 3) assess and
compare available methods for the evaluation of efficacy of
tartar control dentifrices and rinses, including in vitro and
in vivo methodologies. The experimental results are discussed
in terms of our modern concepts for the control of calculus
development.

METHODS AND MATERIALS
Solution/Solid Preparation & Analysis
Solution Preparation/Analysis. Analytical reagent grade chemicals and distilled water were used to prepare solutions and
titrants, which were filtered (0.22 um Millipore) to remove
particulate impurities. Inhibitor solutions were prepared
using reagent grade chemicals, except for disodium ethanehy-
droxy diphosphonic acid (EHDP) which was donated by Norwich
Eaton Pharmaceuticals.
Synthetic Seed Preparation/Analysis. Calcium phosphate seed
materials were characterized by solution chemical, X-ray powder diffraction (XRD) and IR spectroscopy and compared to
literature values summarized recently by LeGeros (39). Specific surface areas (SSA) were measured by single point N_2/He
BET analysis. Synthetic hydroxyapatite (HAP) (Ca/P ratio
1.66/SSA = 28 m^2/g) was prepared by the method of Nancollas
and Mohan (40) and stored at 37°C as an aqueous slurry (46
mg/ml). Octacalcium phosphate (OCP) (Ca/P ratio 1.38/SSA =
29.7 m^2/g) was prepared by the method of LeGeros (41) and
stored dry in a vacuum dessicator before use. DCPD seed material (Ca/P ratio 1.02/SSA = 1.8 m^2/g) was prepared by the
method of Marshall and Nancollas (29) and stored as an aqueous
slurry at pH = 5.35 (50 mg/ml). A high surface area apatite,
DHAP was prepared by a method developed by Lanzalaco and asso-
ciates in our laboratories (42). 50 g of Bio-rad HTP HAP
powder were added to 500 ml of water and acidified to pH 2
with 6 M HCl. This slurry was refluxed for 72 h after which
the pH was adjusted to 12 with KOH. This solution was refluxed for 120 h, the supernate liquid was removed by repeated
decantation and the solid was stored at 37°C in a water slurry
(100 mg/ml). The resulting solid, had a SSA of 51 m^2/g,
Ca/P ratio of 1.61 and XRD and IR patterns resembling poorly
crystalline HAP (39).

Constant Composition (CC) Crystal Growth

CC crystallization experiments followed the general protocol set forth in prior publications (37,38,43-44). The supersaturated reaction solutions each contained a background of NaCl solution required to bring the solution ionic strength to 0.15 M for all studies. The composition of reaction solutions included: 1.0 mM Ca, 0.6 mM P, pH = 7.4 (HAP); 1.75 mM Ca, 2.00 mM P, pH = 7.4 (HAP, DHAP); 5.11 mM Ca, 3.84 mM P, pH = 6.00 (OCP); 10 mM Ca, 10 mM P, pH = 5.55 (DCPD). Following inoculation with seed crystals the reaction solution supersaturation was kept constant by the controlled addition of stock lattice ion titrant solutions from a pair of piston driven burettes as previously described in detail (44). The composition of titrant and reaction solutions (and supersaturation of the media) was determined from mass balance and electroneutrality expressions taking into account the formation of ion pairs (43,45-47). During mineralization, solution aliquots were periodically withdrawn and analyzed to verify constant composition (48). Solid phases were collected and examined by XRD, IR and SEM techniques.

pH Stat Mineralization of DHAP

In these experiments, a pH stat was used to follow the mineralization reaction of DHAP. Supersaturated solutions (Ca 1.75 mM, P 2.0 mM, pH = 7.4, IS = 0.15) were prepared and monitored with a pH electrode. In this case, the decrease in pH associated with crystal growth was compensated for by the addition of 0.01 M KOH as titrant to keep the pH constant during mineralization. Solution calcium and phosphate were again followed by repeated sampling during the experiments as described above.

DCPD Hydrolysis

In these experiments a pH stat followed the hydrolytic conversion of DCPD to apatite-like phases at pH = 8.0. Thermostatted reaction solutions (0.7 mM Ca & P; 0.15 M NaCl; 37°C) were inoculated with 50 mg of DCPD seed.

Topical Inhibitor Effects: Seed Pretreatment Methodologies

Antitartar dentifrices were tested for activity using a mineral seed pretreatment methodology (42). Aliquots of mineral seed slurry were added to 10 ml of dentifrice supernate liquid (10 w/w %). Following 30 seconds vortexing the seeds were centrifuged for 10 minutes at 7500 rpm, the supernatant liquids were poured off, 10 ml of fresh water was added and the seed was again separated by centrifugation. Following the second wash, the seed slurry was filtered (0.22 μm) and the solid was transferred to a 50 ml centrifuge tube containing 5 ml of distilled water. A 2.5 ml aliquot of suspended solid was inoculated into the reaction vessel.

In Vivo Rat Calculus Testing:

Rat calculus testing followed the method of Francis and Briner (49) and was carried out in conjunction with the Oral Health Research Institute of Indiana University. Mixed sex Wistar rats, were balanced into treatment groups of 30. Animals received twice daily treatments (for 5 days a week) of

Table 1
Active/Concentration (μMol/L)

Levels of Inhibitor Actives Effective for 50% Reduction In Mineralization Rate for Calcium Phosphate Phases from Langmuir Inhibition Analysis	Phase	Mg^{2+}	Zn^{2+}	Zn_3Cit_2	PPi	EHDP
	HAP	167	1.7	1.8	2.2	2.4
	OCP	853	0.8	0.8	0.5	0.4
	DCPD	>1000	11.1	15.2	1.4	14.7

1:1 dentifrice:water slurry using a cotton tipped applicator for 30 seconds. Following 3 weeks of treatment animals
were sacrificed and assayed for calculus prevalence. Calculus scores were statistically compared using a Duncans Multiple Range analysis of variance.

RESULTS

Effect of Antitartar Inhibitors on Calcium Phosphate Crystal
Growth- Constant Composition Results
　　　The overall rates of crystallization for HAP, OCP and
DCPD in pure solution measured 3.83 x 10^{-7} mol/min.m^2 (Δ
G = -5.98 kJ/mol), 1.69 x 10^{-6} (ΔG = -1.56), and 8.45 x
10^{-4} (ΔG = -0.98), respectively. These rates, normalized
for specific surface area, showed DCPD to be the fastest
growing mineral phase, consistent with prior observations
(35,36,38). The addition of inhibitors resulted in significant decreases in mineralization rate for all mineral phases.
Calculations demonstrated that the observed reactivity was
not due to solution chelation of lattice ions by inhibitor
molecules but instead pointed to surface adsorption of inhibitors onto growth sites on the mineral surfaces. Inhibition
data were thus analyzed in terms of a Langmuir adsorption
isotherm model, as has been described (33,37,44,50). Plots
of crystal growth rates in this form permitted the direct
comparison of inhibitor efficacy for various phases as shown
in Table 1.
　　　In general, the commercially utilized TC actives similarly decreased crystallization of the three phases of calcium
phosphate at concentrations ranging from 1-100 μmol/L. While
EHDP and pyrophosphate showed similar inhibition for HAP, the
pyrophosphate was superior on DCPD. Magnesium, which is
present in both saliva and mature calculus showed decreased
inhibitory efficacy for OCP and DCPD, while demonstrating
strong inhibitory efficacy for HAP.
Inhibitor Effects on DCPD Hydrolysis
　　　Crystal growth inhibitors also significantly delayed the
hydrolytic transformation of DCPD in aqueous solution. Under
the conditions of these experiments, the onset of DCPD hydrolysis took place in 61 ($\pm$7) minutes. The addition of $ZnCl_2$
and $MgCl_2$ at concentrations of 5 μMol/L had no effect on
the hydrolysis, while both pyrophosphate and EHDP increased
the time needed for initiation of conversion to apatitic
phases by over three fold, at identical concentrations.

Table 2

Dentifrice	% Calculus Reduction
Placebo	----
0.5% Zn_3Cit_2/MFP	26.8
2.0% $ZnCl_2$/NaF	30.6
5.0% PPi/NaF	35.8
2.0% EHDP/NaF	44.7
$P < 0.05$	

Results of Rat Calculus Testing on Antitartar Dentifrices

Effect of TC Dentifrices on HAP

Figure 1 shows the effect of pretreatment of crystalline HAP with antitartar dentifrices. Pretreatment in each case resulted in almost complete initial inhibition of HAP mineralization. Following an induction period (< 1 h for all groups) crystallization commenced and eventually matched non-inhibited levels.

Factors Limiting Inhibitor Efficacy: Overgrowth Experiments

Figure 2 shows the crystallization of OCP for an extended period at sustained supersaturation in the presence of PPi as a crystallization inhibitor. Following an induction period, the crystallization rate increased to levels similar to those in the absence of PPi (similar to the HAP pretreatment results shown above). SEM observations showed the formation of many small crystals having a morphology similar to OCP following this lag period suggesting the secondary nucleation of new crystallites into solution. The addition of extra PPi inhibitor again reduced the mineralization rate, presumably by adsorbing onto these new crystallites and replacing the inhibitor that was depleted by the formation of new crystals. Similar results were obtained with other inhibitors on HAP and DCPD seed materials.

Figure 3 shows the effects of storage time on the inhibitory efficacy of pyrophosphate dentifrices toward DHAP mineralization. As shown, the PPi dentifrice demonstrated significant decreases in efficacy with time. Re-treatment of samples each day restored efficacy, as did the inoculation of mineralization medium with solution PPi.

In Vivo Rat Calculus Studies

Table 2 shows the results of a rat calculus study examining antitartar dentifrices for efficacy.

DISCUSSION

The traditional mechanistic view of antitartar inhibitory species shows these agents coating the surface of calculus crystals during topical exposure, slowing the growth kinetics of these crystals within the plaque. Although the results of the studies reported here support the importance of "inhibition" per se, close examination suggests that the detailed mechanism of inhibitor reactivity is more complex. The Langmuir analysis of CC crystallization rates shows, for example, that saturation coverage of growth sites on the

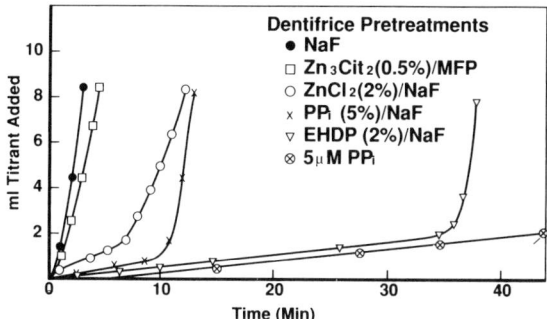

Figure 1.

Effect of pretreatment of HAP with denti-frice supernatants (10 w/w %) on minerali-zation rate during constant composition (Ca 1.75 mMol/L, P 2.00 mMol/L, pH = 7.4, I = 0.15). Data points labelled ⊗ were mineralized without pre-treatment in presence of 5 μMol/L PPi.

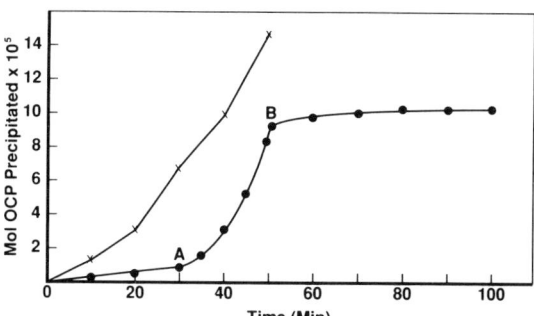

Figure 2.

Overgrowth of OCP in the presence of PPi inhibitor. x control experiment without PPi. ●, experiment initiated with 0.32 μMol/L PPi in solution. Point A shows region where overgrowth initiated via secondary nucleation. Point B denotes the addition of extra PPi, in this case to a solution concen-tration of 3.5 μMol/L.

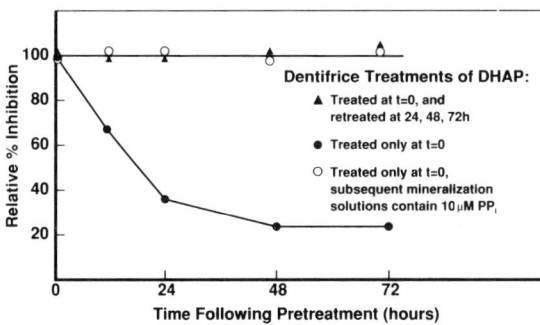

Figure 3.

Effect of storage time following pretreat-ment on the mineralization inhibition effi-cacy of PPi dentifrice (conventional crystal growth of DHAP). Samples were stored in saturated solution between treatments. % inhibition is normalized to experiments per-formed immediately following treatment.

HAP,OCP and DCPD phases occurs at concentrations 2-3 orders of magnitude lower than inhibitor concentrations used in antitartar dentifrices and mouthwashes. (Thus, if we use Zn^{2+} as an example, for HAP and OCP, the concentrations of inhibitor needed for 50 % efficacy ranged from 2.19-0.5 μM/L, in sharp contrast to the 50 mM/L concentration found in sali-va during toothbrushing with a 2 % $ZnCl_2$ toothpaste!) Similar calculations demonstrate that effective TC agents are generally formulated at concentrations well above monolayer coverage level for mineral crystallites. Despite this, the CC studies on pretreatment effects of TC dentifrices shown in

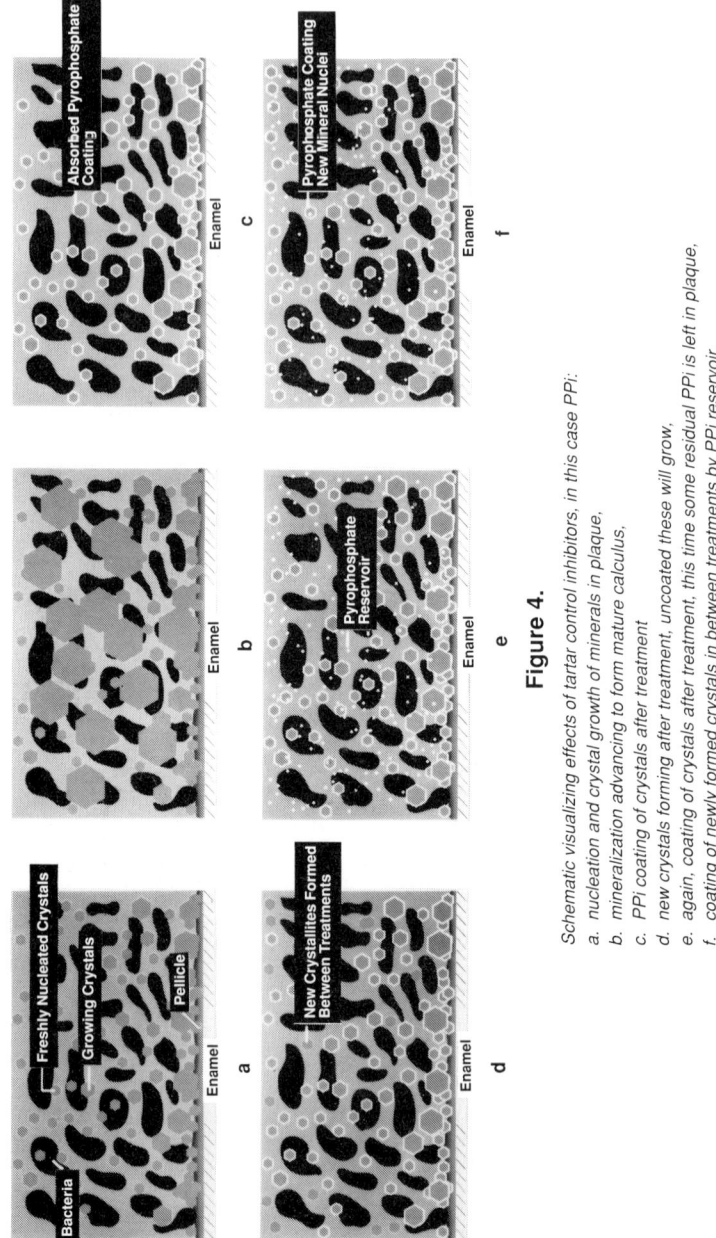

Figure 4.

Schematic visualizing effects of tartar control inhibitors, in this case PPi:

a. nucleation and crystal growth of minerals in plaque,
b. mineralization advancing to form mature calculus,
c. PPi coating of crystals after treatment
d. new crystals forming after treatment, uncoated these will grow,
e. again, coating of crystals after treatment, this time some residual PPi is left in plaque,
f. coating of newly formed crystals in between treatments by PPi reservoir.

Figure 3 demonstrate that adsorption onto existing crystallites provides only temporary protection against mineralization processes in and of itself, with secondary nucleation
and/or overgrowth occurring on even the best inhibitory species. Similarly, the results of studies at low ambient inhibitor concentration show overgrowth which must be compensated
for by extra inhibitor addition in the case of significant
new mineral accumulations (Figure 2). Another factor which
can limit the efficacy of inhibitors is the degradation of
inhibitor molecules, such as the hydrolytic cleavage of pyrophosphate (51-52), shown in Figure 3. Thus, while these
results support the general action of inhibitors, they clearly demonstrate that coating action on growth sites of existing crystals cannot be a sole determinant of efficacy.

A more comprehensive view of inhibitor action is that
inhibition of mineralization within the plaque is due to both
the surface coating of crystals during treatment and the
residual action of inhibitors retained as a "reservoir" within the plaque fluid between treatments. This process is
diagrammatically shown in Figure 4 a-f where the residual
action of tartar control inhibitors is visualized as providing protective coatings for newly developing crystals within
the plaque matrix. The importance of a reservoir source of
inhibitor was previously recognized by Briner and Francis,
who observed significantly greater anticaluluus efficacy for
EHDP when added in low concentrations within the diet (i.e.
simulating low level continuous exposure), as compared with
high concentrations used topically (16). Treatment groups
which received EHDP subcutaneously showed no calculus inhibition, clearly demonstrating that the EHDP action was topical,
rather than systemic (16). Francis postulated that a reservoir of inhibitor agent may present itself as a surface phase
on the existing enamel and tartar crystallites following
treatment with high concentrations of inhibitor molecules
(53-54). In this context, the tartar and enamel crystal
surfaces could be visualized as acting as reservoir sources
of inhibitor through slow release desorption. More recently,
Gaffar et al. (55) suggested that the inhibition of pyrophosphate degradation within plaque (retaining this reservoir) provided added efficacy for calculus inhibition by this
active.

With these thoughts in mind, it is instructive to compare
the levels of inhibitors which must be maintained to achieve
efficacy, in this case 50 % inhibition of mineralization. As
shown in Table 1, these levels are in the µM range for all
actives for OCP and HAP mineralization, but are increased ten
fold for the inhibition of DCPD (with the notable exception
of PPi, which maintained similar efficacy as for HAP). These
findings, suggesting the need for higher levels of ambient
inhibitor for DCPD inhibition may be particularly critical
from two standpoints. First, DCPD is the fastest growing
calcium phosphate phase under physiologic conditions. Thus,
non-inhibited, or only partially inhibited DCPD growth due to
insufficient ambient inhibitor reservoir could result in
large amounts of proportionate mineralization. Secondly,
DCPD has a lower density than HAP by a factor of ⁻25 %. One
could speculate that this mineral would occupy a larger vol-

ume of the plaque thereby setting up a matrix for the transformation to mature tartar. It is noteworthy that pyrophosphate ions maintained efficacy for DCPD mineralization at
10X lower levels than the other inhibitors in these experiments, perhaps explaining the excellent clinical efficacy for
this active despite its known hydrolytic sensitivity. These
cumulative results suggest that longer lasting ambient levels
of inhibitor should have positive clinical effects. In support of this, recent in vitro plaque uptake studies in our
laboratory (56) demonstrated increased PPi uptake and retention in plaque from a dentifrice containing 5.0 % PPi vs 3.3
% PPi (both formulated well over the saturation levels for
mineral crystallites). The positive impact of this extra
reservoir of pyrophosphate was substantiated in clinical
trials showing a 14.5 % relative improvement in clinical
efficacy for the 5.0 % PPi formulation (57).   Overall, the
notion of the need for continuous reactivity of agent is
certainly not a new one to dental research. It is well known
that effective antimicrobials such as chlorhexidine exhibit
strong intraoral retention providing extended plaque inhibition benefits (58). Similarly, fluoride retention in carious
lesions and in plaque is correlated with anticaries properties of this agent (59-60). It is logical that the clinical
efficacy of calculus inhibitor agents require sustained action to limit long term mineralizing effects.

In terms of limitations to inhibitor efficacy, the results of this study demonstrate that secondary nucleation of
new mineral under sustained supersaturation conditions
presents a real limitation to topical inhibitory efficacy
that only reservoir action can overcome. The simple "coating" of tartar crystals during treatment is simply not enough
to confer long term efficacy. Interestingly, secondary nucleation of minerals within tartar has long been considered a
process in calculus development, with the so-called "A" cen-
ters seeding the apparent crystallization of different, rapidly growing "B" sites (2). In addition to overgrowth, the
studies showing hydrolytic degradation of PPi show an additional limitation of this type of active.

A further goal of this research was to examine models
available for the assessment of the crystallization inhibitory reactivity of dentifrice formulations. For in vitro
tests, this type of experimentation is logistically difficult
since the toothpaste slurry and supernatant matrix make the
rapid treatment and separation of crystals difficult. Nevertheless, the application of the CC method to screens of various inhibitor formulations, as shown in Figure 3, is useful
since it gives a measure of overall mineral surface effects
of the formulations in the presence of fluoride and other
formulation components. In in vivo testing, all TC dentifrices exhibited efficacy in the rat calculus model, although the
reactivity of the $ZnCl_2$ dentifrice was decreased relative
to clinical experience with this active (61).   Naturally,
differences in mineral components within rat calculus (mostly
ACP with little DCPD observed) could be responsible for variable sensitivity to inhibitor actives vs the human clinicals. (This, incidentally, would also be a factor in the
case of dog calculus models, since the dog forms calcium

carbonate rather than phosphate deposits (62)). In the rat, the use of effective TC agents results in histological as well as crystalline changes within the deposits (15). Furthermore, in the case of metal ion actives, the influence of combined plaque activity is unclear as is the possibility of forming zinc phosphate directly within the plaque deposits. Thus, while the measurement of prevalence of calculus formation (in the rat) simulates, to a degree, human clinical measures, it appears that non-human _in vivo_ models also provide only empirical data on potential efficacy of TC agents, not significantly different than _in vitro_ crystallization screens, as reported here.

Overall, the results of these studies suggest that the reactivity of inhibitors on mineral surfaces accounts for the anti-calculus action of these actives, as is widely accepted. The results show that all currently used "tartar- control" actives have strong inhibitory effects, but all are prone to overgrowth at sustained supersaturations. Pyrophosphate is unique in its hydrolytic degradation at the apatite surface, however its improved reactivity for DCPD likely compensates, explaining its excellent clinical efficacy. Studies of pretreatment reactivity show that the maintenance of soluble levels of inhibitor is vital to protection against further mineralization for all TC actives. Recent clinical studies show increased efficacy for an increased dosage PPi formulation, with _in vitro_ plaque retention studies suggesting increased reservoir as a possible contributor. Interestingly, the reaction profile for magnesium suggests that the limited reactivity of this salivary ion for DCPD mineralization, at physiologic levels, probably explains its inability to reduce tartar deposition despite its known inhibitory effects on the mature phases of tartar such as HAP (38).

On a final note, it is interesting that current models for inhibitor effects still do not explain the translation of the chemical actions of these molecules and the clinical scores obtained using typical measures of area coverage by calculus on the teeth (63-64). Clinical measures show a logarithmic increase in calculus formation with time, with the areas between teeth and inaccessible to easy brushing forming tartar most rapidly (2,65). Effective clinical agents are typically observed to decrease the "plateau" of this logarithmic rise, with little further accumulation after several months exposure for either actives or controls (see for example (17-19)). Indeed, an EHDP dentifrice exhibited similar efficacy for usage over 18 months, with no changes in calculus composition (i.e. mineral content (19)). Clearly, the inhibition action is not directly related to crystal growth alone, and may be tied to the transformation of faster growing phases or adhesion of new deposits as lamellar structures onto existing tartar. Elucidation of the precise factors leading to efficacy using methods like the highly specific and sensitive CC technique may provide insight into further technologies which may be devoted to calculus prevention. It is hoped that in preventing more calculus, we may reach a threshold level necessary to contribute periodontal

health advantages in addition to the cosmetic advantages
currently provided by these agents (66).

ACKNOWLEDGEMENT
This study was supported in part by a Grant# DE03223 from the
NIDR. The authors wish to also thank Ms. Connie Blanken for
her assistance in the preparation of this manuscript.

REFERENCES
1. Driessens, F.C.M. (1982) in Mineral Aspects of Dentistry, (ed. Myers, H.W.) pp 154-158. Karger, Basel,
   Switzerland.
2. Schroeder, H.E. (1969) Formation and Inhibition of Dental Calculus, 212 pages, Hans Huber, Berne, Switzerland.
3. LeGeros, R.Z., Orly, I., LeGeros, J.P., Gomez, C.,
   Kazimiroff, J., Tarpley, T. and Kerebel, B. (1988) Scanning Microscopy, 2, 345-356.
4. Gron, P., van Campen, G.J. and Lindstrom, I. (1967) Arch.
   Oral Biol., 12, 829-837.
5. Kani, T., Kani, M., Moriwaki, Y. and Doi, Y. (1983) J.
   Dent. Res., 62, 92-95.
6. Friskopp, J. and Hammerstrom, L. (1980) J. Periodontal.,
   51, 553-62.
7. Jenkins, A.T. and Dano, M. (1954) J. Dent. Res., 33, 741-50.
8. Jenkins, A.T. and Dano, M. (1957) Acta Odont. Scand., 16,
   121-26.
9. LeGeros, R.Z. and Shannon, I.L. (1979) J. Dent. Res., 58,
   2371-2377.
10. Mandel, I.D. (1987) Comp. Cont. Educ. Dent. Suppl., 8,
    5235-5241.
11. Shankwalkar, G.B. (1975) J. Ind. Dent. Ass., (special issue)
    pp. 303-310.
12. Weinstein, E. and Mandel, I.D. (1964) J. Oral. Ther. and
    Pharm., 1, 327-334.
13. McNeal, D.R. (1969) J. Public Health Dent., 3(special is-
    sue), 138-152.
14. Ennever, J. and Sturzenberger, O.P. (1961) J.Periodontol.,
    32, 331-333.
15. Francis, M.D. and Briner, W.W. (1973) Calc. Tiss. Res., 11,
    1-9.
16. Briner, W. W. and Francis, M.D. (1973) Calc. Tiss. Res., 11,
    10-22.
17. Sturzenberger, O.P., Swancar, J.R., and Reiter, G. (1971) J.
    Periodont., 42, 416-419.
18. Muhlemann, H.R., Bowles, D., Schait, A., and Bermmaulin,
    J.P. (1970) Helv. Odont. Acta, 14, 31-33.
19. Suomi, J.D., Horowitz, H.S., Barbana, S.P., Spolski, V.W.,
    and Heifetz, S.B. (1974) J.Periodontol., 45, 139-145.
20. Zacherl, W.A., Pfieffer, J.H., and Swancar, J.R. (1985) J.
    Am. Dent. Assoc., 110, 737-738.
21. Mallatt, M.E., Beiswanger, B.B., Stookey, G.K., Swancar,
    J.R., and Hennon, D.K. (1985) J. Dent. Res., 64, 1159-1162.
22. Lobene, R.R. (1986) Clin. Prev. Dent., 8, 5-7.
23. Schiff, T.G. (1987) Clin. Prev. Dent., 9, 13-16.
24. Lobene, R.R., Soparkar, P.M., Newman, M.B., and Kohut, B.E.

(1987) J. Am. Dent. Assoc., 114, 350-352.
25. Stephen, K.W., Burchell, C.K., Huntington, E., Baker, A.G., Russell, J.I., and Creanor, S.L. (1987) Caries Res., 21, 380-84.
26. Lobene, R.R. (1987) Clin. Prev. Dent., 9, 3-8.
27. Francis, M.D., Russell, R.G.G., and Fleisch, H. (1969) Science, 165, 1264-1266.
28. Nancollas, G.H. and Tomazic B. (1974) J. Phys. Chem., 78, 2218-2225.
29. Marshall, R.W. and Nancollas, G.H. (1969) J. Phys. Chem., 73, 3838-3844.
30. Meyer, J.L. and Nancollas, G.H. (1973) Calc. Tiss. Res., 57, 153-161.
31. Posner, A.S., Blumenthal, N.C., and Betts, F. (1984) in Phosphate Minerals (ed. Nriagu, J.O., Moore, P.B.) pp 330-350, Spinger Verlag, N.Y.
32. Nancollas, G.H. (1984) in Phosphate Minerals, (ed. Nriagu, J.O., Moore, P.B.) pp 137-154, Springer-Verlag, N.Y.
33. Nancollas, G.H. (1979) Adv. Coll. Int. Sci., 10, 215-252.
34. Tomson, M.B. and Nancollas, G.H. (1978) Science, 200, 1059-60.
35. Nancollas, G.H., Salimi, M.H., and de Rooij, J.F. (1984) Fortschr. Urol. Nephrol., 22, 198-206.
36. Zawacki, S.J., Koutsoukos, P.B., Salimi, M.H., and Nancollas, G.H. (1986) in Geochemical Processes at Mineral Surfaces (ed. Davis, J.A., Hayes, K.F.) pp 650-662, American Chem. Soc. Symp. Series No. 323.
37. Amjad, Z. (1987) Langmuir, 3, 1063-69.
38. Salimi, M.H., Heughebaert, J.C., and Nancollas, G.H. (1985) Langmuir, 1, 119-122.
39. LeGeros, R.Z. and LeGeros, J.P. (1984) in Phosphate Minerals (ed. Nriagu, J.D., Moore, P.B.) pp 351-385, Spinger-Verlag, N.Y.
40. Nancollas, G.H. and Mohan, M.S. (1970) Arch. Oral. Biol., 15, 731-745.
41. LeGeros, R.Z. (1985) Calc. Tiss. Int., 37, 194-197.
42. Lanzalaco, A.C. and Sunberg, R. (1988) personal communication.
43. Koutsoukos, P., Amjad, Z., Tomson, M.B., and Nancollas, G.H. (1980) J. Amer. Chem. Soc., 102, 1553-1557.
44. Koutsoukos, P. (1980) PhD Thesis, SUNY at Buffalo, 401 pp.
45. Nancollas, G.H. (1966) Interactions in Electrolyte Solutions, pp.73-92, Elsevier, Amsterdam.
46. Davies, C.W. (1962) Ion Association, pp 41, Butterworths, London, UK.
47. Koutsoukos, P.G. and Nancollas, G. H. (1981) J. Dent. Res., 60, 1922-1928.
48. Tomson, M.B., Barone, J.P., and Nancollas, G.H. (1977) Atomic Absorpt. Newsletter, 6, 117-118.
49. Francis, M.D. and Briner, W.W. (1969) J. Dent. Res., 48, 1185-1195.
50. White, D.J., (1982) PhD Thesis, SUNY @ Buffalo 640 pp.
51. Moreno, E.C., Aoba, T., and Margolis, H. (1987) Comp. Cont. Ed. Dent. Suppl., 8, 256-266.
52. Meyer, J.L. and Reddi, A.H. (1985) Arch. Biochem. Biophys., 242, 532-539.

53. Francis, M.D., Gray, J.A., and Greibstein, W.J. (1968) Adv. Oral Biol., 3, 83-120.
54. Francis, M.D., Slough, C.L., Briner, W.W., and Oertel, R.P. (1977) Calc. Tiss. Res., 23, 53-60.
55. Gaffar, A., Polefka, T., Afflito, J., Esposito, A., and Smith, S. (1987) Comp. Cont. Ed. Dent. Suppl., 8, 242-250.
56. Sammons, M.C., Coombs, M.A., Deibel, R.M., Collier, W.G., McCormack, L.D., and White, D.J. (1988) Abstract submitted to 1989 AADR Meeting.
57. Lu, K.H., Ruhlman, C.D., Chung, K., Adams, A., and Bollmer, B., (1988) J. Indiana Dent. Assoc., 67, 17-18.
58. Nikiforuk, G. (1981) in Understanding Dental Caries, pp. 265-267, Karger, Basel, Switzerland.
59. White, D.J. (1988) Caries Res., 22, 27-36.
60. Mellberg, J.R. and Ripa, L.W. (1983) in Fluoride in Preventive Dentistry, 290 pages, Quintessence Publ., Chicago.
61. Rustogi, K.N., Volpe, A.R., and Petrone, M.E. (1988) Comp. Cont. Educ. Dent., 9, 78-79.
62. LeGeros, R.Z. and Shannon, I.L. (1979) J. Dent. Res., 58, 2371-2377.
63. Volpe, A.R., Manhold, J.H., and Hazen, S.P., (1965) J. Periodontol., 36, 299-304.
64. Volpe, A.R., Kupczak, L.J., and King, W.J. (1967) Periodontics, 5, 184-193.
65. Conroy, C.W. and Sturzenberger, O.P. (1968) J. Periodontol., 39, 20-22.
66. Mandel, I.D. and Gaffer, A. (1987) J. Clin. Periodontol., 13, 249-257.

J.Arends
A.G.Dijkman
J.Ruben
W.L.Jongebloed[1]

The role of diphosphonates on calculus formation and remineralization

Laboratory for Materia Technica, University of
Groningen, Antonius Deusinglaan 1, 9713 AV
Groningen, The Netherlands and [1]Department of
Histology and Cell Biology, Oostersingel 69/2,
9713 EZ Groningen, The Netherlands

ABSTRACT

Dental calculus is a deposition of mainly hydroxyapatite
(HAP) on the pellicle, whereas remineralization is a
precipitation of HAP in enamel. The actual distance between
calculus sites and positions of remineralization is very small.
Diphosphonates influence mineralization of plaque and might
influence remineralization. The results of this work show that:
lesions formed in vitro in the presence of MHDP, do
remineralize substantially in vivo as well as in vitro. MHDP
inhibits mineral deposition from saturated HAP solutions in
vitro, with respect to controls. The effect of MHDP on
fractured calculus is such that in <u>part</u> of the calculus HAP
precipitation/growth is prevented.

Recent Advances in the Study of
Dental Calculus

INTRODUCTION

Dental calculus forms supra- and subgingivally resulting from mineralization of the bacterial plaque adsorbed onto the tooth surfaces. Most emphasis is on supragingival calculus. Firstly, because it forms much faster than subgingival calculi (1), and secondly because presently several anticalculus dentifrices significantly reduce new supragingival calculus deposits after oral prophylaxis. Dental calculus consists of a large number of inorganic crystalline components, organic components and some unmineralized plaque on the surface.

Dental calculus is as far as composition is concerned, the most complex mineralized tissue in humans. The mineral part of calculus 70-90 wt%, consists of hydroxyapatite (HAP), DCPD (dicalciumphosphate-dihydrate), OCP (octacalciumphosphate), whitlockite and other minerals. About two third of the inorganic component is crystalline (2). HAP is the main mineral phase found in all calculus types and is formed at all ages.

HAP is also the main mineral component in enamel, dentine and cementum. Therefore, agents preventing or inhibiting calculus formation might as well influence mineral processes near the tooth surface e.g. remineralization. Calculus formation is a deposition of HAP (mainly) <u>on</u> the pellicle of the teeth, whereas remineralization takes place under the pellicle <u>in</u> the teeth. Starting with a cleaned tooth surface, the distance between areas where calculus is situated and where remineralization takes place is very small; most likely in the order of 1 μm.

The prevention of calculus formation by mechanical scaling or mechanical cleaning with a toothbrush does from the mechanistic point of view, not influence remineralization. In attempts to prevent calculus formation by means of mineral inhibition, numerous substances have been added to toothpastes, rinses etc. The agents used have been selected either because they decrease dental plaque formation or because the agents influence the plaque mineralization. They could influence mineralization phenomena in teeth as well.

Calculus formation has been effectively prevented in man and animals by the topical application of diphosphonates (3-6). Diphosphonates are a family of compounds interacting strongly with minerals (7) and are thought to prevent calculus deposition by inhibiting HAP crystal growth.

Aims of this investigation were threefold:

I.   to compare in vivo and in vitro remineralization of lesions formed in human enamel in the presence of diphosphonate.

II. to investigate mineral precipitation on enamel in vitro from supersaturated calciumphosphate solutions.

III. to study mineral precipitation in vitro from supersaturated calciumphosphate on natural calculus pretreated with diphosphonate solutions.

MATERIALS AND METHODS

Freshly extracted bovine teeth were employed. The labial surfaces were partially ground flat on abrasive paper (Siawat

600) and water. The enamel surrounding the flat area was cut away using a water-cooled diamond saw, in such a way that the final polished enamel sample had a surface area about 3 x 4 mm. Artificial initial enamel lesions were produced in a solution containing 3 mM $CaCl_2$. $2H_2O$, 3 mM KH_2PO_4, 50 mM CH_3COOH and 6 µM MHDP (or 10 µM F^- as NaF) in a constant composition set-up for 140 h. The constant composition apparatus has been described previously (8). Twentyfour teeth were demineralized as described with MHDP in solution. 12 were remineralized in vitro; 12 in the in vivo experiment. The remineralization in vitro was done in the constant composition apparatus using a solution of 1.5 mM $Ca(NO_3)_2$, 0.9 mM KH_2PO_4, 20 mM HEPES buffer at pH = 7.0 for 14 days.

For the in vivo experiment 6 participants carried 12 samples in the buccal flanges of a full prothesis for 4 weeks. The enamel surface was flush with the outer surface of prosthesis (9). Toothbrushing was done with a 1000 ppm F toothpaste twice a day. No denture cleaners were used. The patients carried the prosthesis day and night.

Analysis of all samples was done by microradiography using the technique previously published (10). Both mineral loss ΔZ in vol% µm and lesion depth l in µm were determined. To obtain adequate controls, 2 sections of the demineralized enamel were removed after in vitro demineralization but before remineralization (in vitro or in vivo). The remineralization data were obtained from 3-5 sections of each tooth.

The precipitation on surfaces from saturated HAP solutions - either calculus or enamel- was studied in vitro. These solutions were saturated for more than 1 year and had a pH = 7.0. Two groups of 6 enamel samples were stored in these solutions at 37°C for 2 weeks. Group A was pretreated with 6 µM MHDP for 1 h and stored; Group B was stored only. After washing the samples were sputter-coated with gold and observed in a JEOL 35C SEM.

Calculus was collected from patients between 40-50 y of age and stored at 100% RH. All samples were fractured several times and 3 groups of 8 samples formed. Group 1 was untreated; Group 2 was pretreated with 6 µM MHDP for 1 h and stored in the saturated HAP solution at 37°C for 2 weeks. Group 3 was stored in the HAP solution only. SEM observation was done on the FRACTURED surfaces as described.

RESULTS

The results of the in vitro and in vivo remineralization experiment are compiled in Table 1. The in vitro data of ΔZ and l are presented before and after remineralization if the lesion was formed in the presence of 6 µM diphosphonate or 10 µM F^-. In vivo remineralization data are presented using the procedure discussed extensively in Dijkman et al. (9). Participants in which the lesion depth decreases as compared to the lesion depth at the beginning of the remineralization experiment ("remineralizers") and participants with an increasing lesion depth during the in vivo period, have been separated.

J.Arends *et al.*

Table I. Mineral loss (ΔZ) and lesion depth (l) values of the remineralization experiments in vitro and in vivo in lesions demineralized in the presence of 6 µM MHDP or 10 µM F⁻. ΔZ values are in vol% mineral x µm; l values are in µm. Mean values ± standard deviations are presented. n is the number of specimens. The remineralization period is in days.

after		in demin. solution	period of remin.	l	ΔZ	n
in vitro	demin.	6 µM MHDP	-	120±25	3100± 900	12
in vitro	remin.		14 d	80±20	2100± 700	12
in vitro	demin.	10 µM F⁻	-	98±20	1700± 200	12
in vitro	remin.		14 d	negligable	negligable	12
in vitro	demin.	6 µM MDHP	-	110±15	3000± 800	12
in vivo	remin.		30 d	50±15	1500± 400	10
in vivo	remin.		30 d	130±20	3800±1000	2

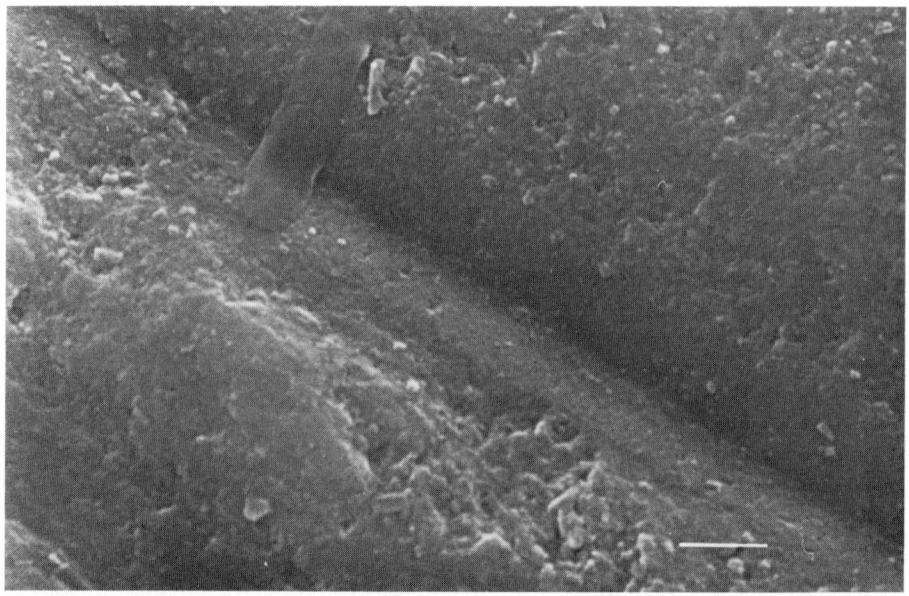

Figure 1A. Crystalline precipitates from saturated HAP on enamel; The bar is 1 µm.

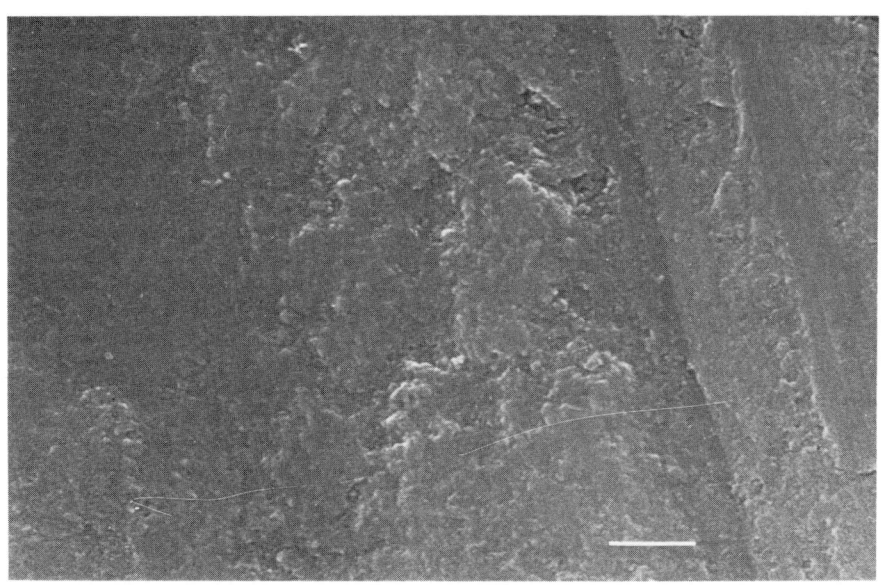

Figure 1B. As Fig. 1A but after MHDP pretreatment. No or very little crystalline precipitation on the surface. The bar is 1 µm.

Figure 2. Calculus fractured only; the bar is 20 µm.

Enamel subjected to saturated HAP solutions behaves quite different if the tissue was MHDP pretreated or not. Figs. 1A and 1B show representative micrographs at about 10,000x. The untreated enamel surface (Fig. 1A) shows crystal formation on the surface; the precipitates are presumed to be calciumphosphates. The precipitates are not observable in Fig. 1B where the enamel was MHDP pretreated.

Calculus FRACTURED and subsequently stored in saturated HAP solutions is shown in Figs. 2 to 4. Fig. 2 shows a representative control; various agglomerated crystal structures, bacteria and "porosities" are noticable. Fig. 3 illustrates a similar sample but after storage in the HAP solution. Crystalline precipitates are present everywhere (see arrows) over large areas. Fig. 4 shows a similar sample pretreated with MHDP and then stored in HAP. Crystalline precipitates are visibly locally (arrows) but not in neigbouring areas. Crystals form on particular spots only; not over large surface areas.

DISCUSSION

The remineralization data in vivo and in vitro compiled in Table 1 show that lesions formed in the presence of MHDP do remineralize in vivo as well as in vitro. In 2 weeks, lesions formed in the presence of MHDP were reduced in lesion depth and mineral loss values by 33% and 32%, respectively (in vitro). During one month in vivo remineralization similar lesions were reduced in lesion depth and mineral loss by 54% and 50% resp., for 5 of the 6 participants. One participant did not remineralize at all but continued to demineralize even in plaque free conditions (see also Dijkman et al. (9)). This is in agreement with previous observations that about 15-20% of all patients do no noticably remineralize in intra oral experiments.

The remineralization capacity of lesions formed in the presence of MHDP is, however, in vitro seriously impaired, if compared to in vitro lesions formed in the presence of F. The last mentioned lesions remineralize nearly completely (see Table 1).

The SEM data in Fig. 1 show crystal formation/precipitation in saturated HAP solutions on enamel surfaces when no MHDP pretreatment was given. (this was found for enamel with a pellicle and for polished samples).

Combining the above results one can draw the conclusion that MHDP present on enamel inhibits growth on the enamel surface, but also that the MHDP present inside the lesion does not strongly influence in vivo remineralization.

The effect of MHDP on calculus indicates that on fractured surfaces, the inhibitor influences some parts of the calculus in such a way that further growth seems to be prevented. This is, however, only true for PART of the fractured calculus surface. Presently it is not known why part of the fractured calculus surface behaves different from others. Speculating this might be due to a difference in chemical phases, difference in chemical composition, or due to differences in organic components.

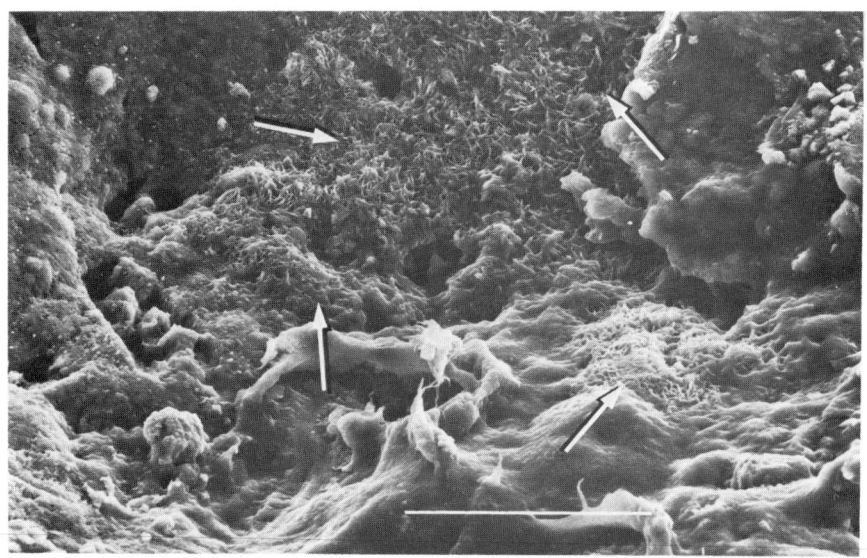

Figure 3. Fractured calculus stored in saturated HAP. Crystalline precipitates are noticable over large areas (arrows); the bar is 100 μm.

Figure 4. Fractured calculus MHDP pretreated and subsequently stored in saturated HAP solutions. Crystalline precipitates are observable but only locally (arrows); the bar is 100 μm.

In this study a static situation of MHDP containing enamel, MHDP containing surfaces and MHDP containing calculus subjected to remineralizing/precipitating conditions is described. In these static conditions small amounts of MHDP present seem to be beneficial against surface precipitations, against further calculus growth and not really influencing remineralization. Under dynamic conditions e.g. daily diphosphonate use the results might be different. In conclusion diphosphonates having numerous properties similar to pyrophosphate, are interesting compounds to consider as anti-calculus agents.

REFERENCES

1. Mandel, I.D. (1967) J. of Periodont., 38, 721-732.
2. Little, M.F. (1963) J. of Dent. Res., 42, 78-83.
3. Briner, W.W. and Francis, M.D. (1973) Calcif. Tissue Res., 11, 10-22.
4. Francis, M.D. and Briner, W.W. (1973) Calcif. Tissue Res., 11, 1-9.
5. Mühlemann, H.R., Bowles, D., Schait, A., Bernimoulin, J.D. (1970) Helv. Odont. Acta, 14, 31-33.
6. Sturzenberger, O.P., Swankar, J.R. and Reiter, G. (1971) J. of Periodont., 42, 416-419.
7. Fleisch, H. and Russel, R.G.G. (1972) J. of Dent. Res., 51, 324-332.
8. Buskes, J.A.K.M., Christoffersen, J. and Arends, J. (1985) Caries Res., 19, 490-496.
9. Dijkman, A.G., Schuthof, J. and Arends, J. (1986) Caries Res., 20, 202-208.
10. Ogaard, B., Arends, J., Schuthof, J. and Rölla, G. (1986) Caries Res., 20, 270-277.

J.D.B.Featherstone[1]
D.J.White[2]
W.D.Bowman[2]
M.Shariati[1]
G.K.Stookey[3]
M.Lo Re[4]
G.H.Nancollas[4]

Effects of anticalculus dentifrices on caries processes: *in vitro* and *in vivo* studies

[1]Eastman Dental Center, Rochester, NY, [2]The Procter & Gamble Co., Cincinnati, OH, [3]Indiana University OHRI, Indianapolis, IN, and [4]SUNY at Buffalo, Buffalo, NY, USA

ABSTRACT
The anticalculus efficacy of commercial dentifrice (toothpaste) and mouthrinse formulations results from inhibition of mineralization by key product ingredients, such as pyrophosphate (PPi), zinc chloride, and zinc citrate on pre-calculus plaques and calculus mineral deposits. Since the action of these agents is not necessarily restricted to plaque and calculus, effects on the enamel surface and on subsurface caries lesion reactivity are possible. The latter possibility must be carefully considered in the assessment of overall caries reactivity for anticalculus formulations. This paper reports the results of in vitro pH cycling (caries progression), in vivo rat caries, and constant composition demineralization studies using anticalculus dentifrices containing 0.243 percent sodium fluoride (NaF) and a range of pyrophosphate concentrations. The inclusion of soluble pyrophosphate (3.3, 4.0 or 5.0 percent) in dentifrices containing 0.243 percent NaF had no significant effect on anticaries efficacy as measured by both rat caries and in vitro lesion progression models. The treatment of caries lesions with a 5 percent PPi/0.243 percent NaF dentifrice in the constant composition model resulted in a significant decrease in demineralization kinetics compared with treatment with a NaF dentifrice containing no pyrophosphate. We conclude that the combined effects of agents on de- as well as remineralization must be considered in evaluating the cariostatic reactivity of anticalculus formulations.

Recent Advances in the Study of
Dental Calculus

INTRODUCTION

Toothpaste formulations containing mixtures of soluble pyrophosphate salts have been shown in several clinical trials to be effective for the reduction of dental calculus (1-11). The mechanism behind the calculus efficacy of these formulations involves the adsorption of inorganic pyrophosphate (PPi) onto mineral surfaces in the plaque matrix, slowing down the rates of crystal growth and transformations of calcium phosphate phases (12-17). While mineralization inhibition provides an effective means for calculus control, all inhibitors of calcium phosphate crystal growth have the potential to negatively influence subsurface remineralization of caries lesions. This potential was demonstrated in previous research by Briner et al. (18), Muhlemann and Aeschbacher (19), and ten Cate et al. (20) showing that strong mineralization inhibitors, like ethane hydroxy diphosphonate (EHDP) can under some conditions slow natural maturation and remineralization processes of enamel and incipient caries lesions. Similarly, Regolati and Hotz (21) demonstrated that both EHDP and PPi compete with fluoride (F) for adsorption onto powdered hydroxyapatite.

While the potential effects of anticalculus agents on remineralization processes are a cause for legitimate concern, careful and thorough examination of the anticaries benefits of PPi/NaF dentifrice formulations have proven that these systems exhibit pronounced anticaries reactivity. Recent clinical studies have shown that dentifrices containing 3.3 percent soluble PPi and 0.243 percent sodium fluoride (1100 ppm F as NaF) are effective for the prevention of caries (22-23), confirming the conclusions reached by in vivo remineralization (24), in vitro caries progression (25) and in vitro F uptake (26) studies. Hypotheses advanced to explain the "non-interference" of pyrophosphate with fluoride have included the possible in situ hydrolysis of PPi (24) and potential cooperative PPi + F effects on acid resistance of treated enamel (25). Our present understanding of the anticaries action of fluoride would suggest that the remineralization/demineralization balance would have to be aided or at least maintained in order for an inhibitor not to negatively influence caries formation (27-29). Consequently if the possible beneficial or detrimental effects of anticalculus agents are to be comprehensively assessed in combination with fluoride then profiles must include models where both de- and remineralization are examined.

One of the factors that may affect the overall impact of crystal growth inhibitor/fluoride combinations on caries processes is the ratio of the agents within a formulation, since excess effects of either agent might compromise the reactivity (anticalculus or anticaries) of the other. Recently, an antitartar dentifrice was introduced which contained elevated levels of pyrophosphate (5.0 percent soluble PPi salts) in combination with 0.243 percent NaF, compared to the 3.3 percent PPi formulations previously tested in profile and clinical studies (10). The purpose of the present study was to determine the effects of PPi dose in dentifrices containing 0.243 percent NaF using both in vitro pH cycling and in vivo rat caries models. In addition, the effects of the 5 percent PPi/0.243 percent NaF dentifrice on acid demineralization kinetics of dental enamel were evaluated in a constant composition demineralization system in order to gain further insight into the processes contributing to the overall cariostatic reactivity of these formulations.

METHODS AND MATERIALS

Dentifrice treatment groups

Three dentifrices containing soluble pyrophosphate (3.3, 4.0, 5.0 percent PPi) in combination with 0.243 percent NaF (1100 ppm F) were tested. A dentifrice with 0.243 percent NaF only and a placebo dentifrice with no added NaF or PPi were used as controls. All dentifrices were formulated with a silica abrasive. Dentifrice supernatant liquids, separated by centrifugation of 25 percent (w/w) water slurries, were diluted into TISAB and analyzed for available fluoride using an ion specific electrode.

In vitro pH-cycling de- and remineralization

The pH cycling model of ten Cate and Duijsters (30) was modified as described previously (31) to simulate one month of in vivo demineralization around orthodontic appliances (32). Each test cell consisted of ten human tooth crowns with two exposed windows each [designated upper (toward occlusal surface) and lower (toward cervical margin) respectively]. The teeth were prepared from human premolars with caries-free crowns as previously described (31). The crowns were removed from the roots, brushed with warm detergent solution, rinsed in deionized water and air dried. Tooth crowns were coated with acid resistant varnish, leaving two narrow windows exposed. During all dipping and immersion procedures of the test regime, the windows were readily available to demineralization/remineralization fluids and test slurries. The test regime in each 24 h period included: 1) 6 h demineralization at 37°C in a solution containing 2 mmol/L calcium and phosphate, 0.075 mol/L acetate at pH 4.3, where each tooth crown was immersed individually in 40 mL of solution; 2) 5 minute immersion (following a deionized water wash) in 4 mL slurries of test toothpastes (prepared as 1:3 slurries in deionized water); and 3) 17 h immersion (again following deionized water wash) each in 20 mL of a mineralizing solution composed of 1.5 mmol/L calcium, 0.9 mmol/L phosphate, 0.15 mol/L KCl and cacodylate buffer to pH 7.0 (20 mmol/L) to simulate the natural remineralization potential of saliva (25,27,30-31). One further group was included with no dentifrice treatment as a demineralization/remineralization only control. At the conclusion of the test the tooth crowns were thoroughly rinsed, sectioned longitudinally through the lesions and embedded in epoxy resin with the cut face exposed as reported previously (27,31). This method is an improved version of that originally reported by Purdell-Lewis et al. (33). After serial polishing, each lesion was assessed by microhardness examination according to the techniques in the references cited above, starting at 25 μm from the anatomical tooth surface and progressing inward in 25 μm steps to a depth 300 μm into the tooth, across the sectioned lesion on a line perpendicular to the surface and into the underlying enamel. The indentation lengths were converted via Knoop hardness number to volume percent mineral according to the formula in our previous publications (34,35). Values of volume percent mineral in the underlying sound enamel were normalized to an arbitrary mean of 85 vol. percent mineral in a manner similar to that used for quantitative microradiography. The overall mineral loss (ΔZ) from each lesion was calculated from the data using curve fitting with Simpson's rule (36) to provide values of ΔZ for each lesion and subsequently each group in units of vol. percent x μm.

Rat caries experiments

The study design followed FDA method 37 and was carried out at the University of Indiana OHRI (37). Weanling mixed sex Sprague-Dawley rats (n=150) were received as dams with 7-day old pups (10/litter). Animals were provided with diet MIT-305 and water ad libitum. On day 15 animals received an oral inoculation (0.2 mL culture/animal) of streptomycin resistant S. mutans 6715. On day 18, the animals were weaned with dams rotated among litters. At this point, animals were provided with cariogenic MIT-200 diet and were re-inoculated with S. mutans for three more consecutive days prior to the study initiation. This inoculation included both oral inoculation of occlusal surfaces along with spraying of culture on bedding and addition of 10 mL of culture to the animals' drinking water. On day 20, the animals were given a unique number by ear punch with records kept of littermates, weight and sex, the parameters for stratification. On day 22 the animals were stratified into groups of 30, and housed in pairs in suspended wire-bottomed cages. The cages were arranged so that all animals of the individual groups were together and the cages were labelled with a group designation and treatment. Treatments were initiated on day 22 and consisted of twice daily swabbing of the mandibular and maxillary molars for 30 seconds (15 seconds each side of the mouth) with a cotton tipped applicator which had been dipped into toothpaste slurries (50 percent dentifrice/distilled water). Treatments took place on Monday-Friday (no treatments on weekends) and continued for three weeks. In addition to the treatments, the S. mutans infection in the animals was confirmed on day 22 and the final day of the study by culturing of swabs taken from each animal. At the conclusion of treatments the animals were sacrificed and the teeth were prepared and analyzed as previously described in detail by Keyes (38) and Navia (39).

Constant composition demineralization

Early caries-like lesions were prepared in human enamel using the methods described by White (40), and immersed for thirty minutes in 25 (w/w) percent deionized water slurries of placebo dentifrice (no added fluoride), or 0.243 percent NaF dentifrice (Advanced Formula Crest), or 0.243 percent NaF/5.0 percent PPi (Crest Tartar Control) dentifrice, respectively. Following treatment, specimens were rinsed and placed into a lactate buffer constant composition demineralization system as previously described (41). The rate of lesion progression was monitored under constant solution conditions of 3.5 mmol/L Ca, Ca/P (mol ratio) = 1.67, ionic strength 0.15 mol/L, lactic acid 0.05 mol/L, pH = 4.5.

RESULTS

In vitro lesion progression (pH cycling model)

Figure 1 shows the mean mineral density profiles for each of the dentifrice groups of the study in the lower treatment window. The placebo dentifrice group and the demineralization/remineralization control group showed marked demineralization with the mineral content showing a minimum of 20 vol percent mineral and 35 vol percent mineral respectively, at 25-50 μm depth. The lesion depths for the groups were 125-150 μm. In contrast, all four fluoride-containing dentifrice groups showed minimal demineralization under these test conditions, with only minor (less than 10 vol percent) mineral loss being observed in the outer 50 μm. Similar results were obtained for the upper windows of the teeth. Table I presents the results of the pH cycling when calculated in terms of mineral loss, ΔZ, for both upper and lower windows. These ΔZ values are analogous to those obtained by microradiography (36). Statistical analyses (ANOVA) demonstrated that all four fluoride-containing dentifrices, including the NaF-only positive control and the three pyrophosphate-containing formulations were not significantly different in their ability to diminish caries progression ($p < 0.05$). The placebo dentifrice, on the other hand, permitted dramatic mineral loss not significantly different ($p < 0.05$) from the demineralization/remineralization only control.

In vivo rat caries tests

Table II shows the results of the rat caries test for the five dentifrices, presented as total caries score and the percentage reduction in caries vs the control (placebo) dentifrice group. The four fluoridecontaining dentifrices reduced overall caries formation between 53 and 62 percent compared to the placebo dentifrice treatment. No significant differences ($p < 0.05$ ANOVA) were observed among the four fluoride-containing dentifrice treatments. These were all statistically different from the placebo dentifrice.

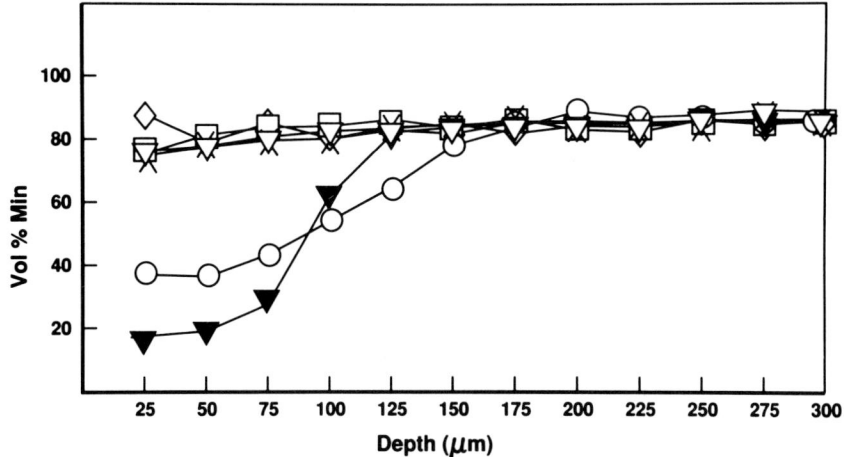

Figure 1. *Lesion progression for dentifrice treatment groups in the lower window:* ◇, *0.243% NaF;* □, *3.3% PPi/0.243% NaF; X, 5.0% PPi/0.243% NaF;* ▽, *4.0% PPi/0.243% NaF;* ▼, *placebo;* ○, *Control.*

Table I. MEAN RELATIVE MINERAL LOSS (ΔZ, μm x VOL PERCENT) FOR UPPER AND LOWER WINDOW GROUPS IN THE pH CYCLING MODEL.

Treatment group	Upper Window		Lower Window	
	ΔZ	SE**	ΔZ	SE
5.0 PPi/0.243 NaF*	572⎤	122	442⎤	196
4.0 PPi/0.243 NaF	270	212	719	243
3.3 PPi/0.243 NaF	465	167	306	139
0.243 NaF (Crest)	785⎦	512	433⎦	235
Placebo	3213⎤	337	4531⎤	414
Demin/remin only	2806⎦	558	4170⎦	735

*   5.0 PPi = 5.0 percent soluble pyrophosphate
     0.243 NaF = 0.243 percent sodium fluoride (1100 ppm F)
** SE = Standard error of the mean. Brackets indicate test groups not
     significantly different (Least significant difference comparison) at
     p < 0.05 (ANOVA).

In vitro demineralization - constant composition studies
    Figure 2 shows the effect of pretreatment of artificial caries lesions
with dentifrice slurries on the demineralization rate during a secondary acid
attack, in this case under conditions of constant demineralization. Treatment
with NaF dentifrice decreased the secondary demineralization rate by 68.4 percent compared to the placebo treated lesions (rates = 2.36×10^{-8} mol min^{-1}
cm^{-2} for placebo-treated and 0.75×10^{-8} mol min^{-1}cm^{-2} for NaF-treated
lesions). Treatment with the combination soluble 5.0 percent PPi/NaF dentifrice resulted in a decrease of 78.9 percent in the overall demineralization

J.D.B.Featherstone *et al.*

Table II.   MEAN CARIES SCORES AND PERCENT REDUCTION VERSUS PLACEBO IN
             THE RAT CARIES MODEL

| | Total Caries Score | | Percent reduction |
Treatment group	Mean	SE**	versus placebo
5.0 PPi/0.243 NaF*	13.6 ⎤	1.9	62 ⎤
4.0 PPi/0.243 NaF	14.0 ⎥	1.9	61 ⎥
3.3 PPi/0.243 NaF	16.9 ⎥	2.5	53 ⎥
0.243 NaF (Crest)	15.1 ⎦	2.4	58 ⎦
Placebo	35.6	4.2	--

*   5.0 PPi = 5.0 percent soluble pyrophosphate
      0.243 NaF = 0.243 percent sodium fluoride (1100 ppm F)
**   SE = Standard error of the mean. Brackets indicate test groups not signif-
      icantly different (Least significant difference comparison) at
      $p < 0.05$ (ANOVA).

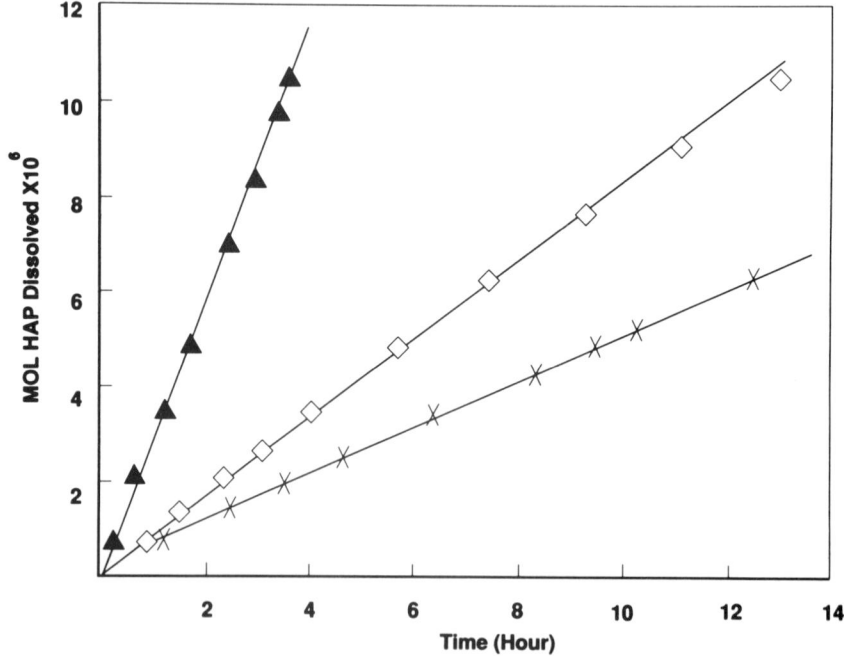

Figure 2.   *Constant composition demineralization
of artificial carious lesions following
dentifrice treatment:* ▲ *, placebo;* ◇ *, 0.243%
NaF; X, 5%PPi / 0.243% NaF.*

rate relative to the placebo (rate = 0.50×10^{-8} mol min^{-1} cm^{-2}) and
33.3 percent decrease relative to the NaF dentifrice-treated lesions.

DISCUSSION

Fluoride is an effective decay preventive agent when added to drinking water or when delivered in topical forms such as commercially available toothpastes and mouthwashes or professionally applied treatments (42). The mechanism behind the decay preventive efficacy of fluoride is believed to involve its strong effects upon both demineralization and remineralization processes in tooth enamel (27-29). Fluoride accelerates the crystal growth rate of apatitic calcium phosphate forming fluorapatite which is a mineral phase of very low solubility and which is thus highly resistant to acid dissolution (29,43,44). Macroscopically, these benefits at enamel crystal surfaces contribute to increased rates of caries lesion repair, through remineralization as well as increased resistance to lesion progression (27-29). From a cariostatic point of view, fluoride delivered to the tooth surface can be viewed as the "ultimate" anticaries agent - enhancing remineralization while simultaneously inhibiting demineralization.

Crystal growth inhibitors, including EHDP, PPi, zinc ion and other anticalculus actives also exhibit strong affinity for enamel mineral crystals. These agents typically decrease both the crystal growth rate and the dissolution rate of chemisorbed surfaces (45-47). The decrease in dissolution rate, as in the case of fluoride, could provide cariostatic benefits to treated teeth. Indeed, many early studies with mineral inhibitory agents like EHDP or pyrophosphate were designed to investigate possible cariostatic benefits of these agents (48). However, in contrast to fluoride, mineral inhibitory agents also restrict the crystal growth processes which are responsible for remineralization. Thus, while fluoride provides positive benefits for both de- and remineralization, crystal growth inhibitors in theory could provide positive benefits only as demineralization inhibitors, with the possibility of negatively influencing remineralization. It is perhaps these contrasting inhibitor effects which limit the capabilities of these molecules to be effective cariostatic agents on their own.

Currently marketed anticalculus toothpastes do not include crystal growth modifiers alone, but also contain some form of fluoride as an anticaries agent. The cariostatic action of these formulations will thus result from the sum of contributions of these two types of components, taking into account any competition that exists, along with the secondary effects that the oral environment might have on each of these species (e.g. the EHDP in vivo interactions reported by Arends et al. in this issue (49)). Obviously, tests to measure the overall cariostatic potential of such mixed systems must include the combination of demineralization and remineralization processes, since individual measurements in a single remineralization or demineralization type protocol might not be predictive of total efficacy.

In the case of PPi/NaF systems, our research groups and others have developed comprehensive profile methodologies which appear to give an accurate reflection of the anticaries activity of these formulations as established by clinical caries studies (22,23). These profile tests include (i) fluoride uptake, (ii) bioavailability (lack of complexation) of fluoride (26), (iii) in vitro lesion progression studies, which utilize the combination of both de- and remineralization (25), and (iv) in vivo models, including animal caries methods (37) as well as in situ denture chip panels (24,50).

In the present study the effects of increased dosages of soluble PPi in a NaF dentifrice were examined by in vitro caries progression and animal caries methodologies. In both protocols, soluble PPi, in the dosage range from 3.3-5.0 percent, had no deleterious effect on the cariostatic performance of the NaF formulations. These results are in agreement with fluoride uptake studies of these same formulations (51).

The constant composition demineralization results reported here showed a dramatic reduction in demineralization following the treatment of the artificial lesions with the 0.243 percent NaF dentifrice compared to placebo-treated lesions and this reduction was markedly enhanced with the 5.0 percent soluble PPi/NaF dentifrice. Similar pronounced effects have also been observed by White and McClanahan (52) in dissolution studies of synthetic hydroxyapatite where PPi/NaF combination treatments gave decreased dissolution rates compared to apatite treated with NaF or PPi only. From a mechanistic standpoint these results are consistent with previous studies where fluoride addition to other surface-active inhibitors in the demineralization media resulted in decreased overall demineralization (53-56). The surface protective effects of inhibitors such as polyacrylic acid, proteins and diphosphonate (MHDP) were magnified by the addition of fluoride. Interestingly, the demineralization retarding effects of the inhibitors in these combination systems appeared to be manifested by their protection of the surface of the caries lesions, resulting in a decreased surface etching as well as more sound outer layers of the prepared lesions (53). Fluoride, on the other hand, strongly influenced demineralization throughout the lesion with increased thickness of the surface zones as well as decreased lesion depth and mineral loss (53,54,56-58). While these differences are in part influenced by the relative mobility of fluoride and inhibitors within the lesions, with fluoride likely diffusing more rapidly than most inhibitory molecules (59), there are also differences in the mechanism of the demineralization inhibitory action of fluoride and other inhibitor molecules (e.g. pyrophosphate) at the molecular level.

The effect of fluoride on apatite dissolution may stem from its substitution in the hydroxyapatite (HAP) lattice, forming fluorapatite (FAP) mineral (29,43-44). The site of fluoride adsorption, or more correctly its substitution, is the OH site within the apatite. Traditionally, the decreased thermodynamic "solubility" of the FAP mineral (with a Ksp two orders of magnitude less than HAP) is viewed as the reason why fluoride increases resistance to dissolution. However, as Brown et al. have pointed out (43) the pronounced effect of fluoride on enamel demineralization must be due to more than simple solubility product differences between HAP and FAP. Recent research has shown that the local effects of fluoride present in the solution phase within the "partially closed" system of carious enamel may be much more important than the effects of fluoride bound within the mineral crystals (29,43-44, 60-61). Fluoride affects dissolution rates in a "thermodynamic" manner since fluoride in solution (which can be released from the crystal) changes the solution undersaturation with respect to the partially substituted mineral.

In contrast to this effect of F, other anionic inhibitors, such as PPi, do not readily substitute in the calcium apatite lattice nor do they affect the thermodynamic properties of solutions (at these low levels) during dissolution. These molecules chemisorb onto the surface of the apatite crystals covering active sites for the propagation of dissolution, including holes and kink sites, thereby slowing the dissolution rate in a 'kinetic' manner. Importantly, the release of these molecules into solution has little or no effect on the undersaturation of the demineralization media. Thus, the net effect on dissolution must come from surface coverage effects at the mineral interface. The influence of these inhibitor species on cariostasis could resemble the action of the natural salivary pellicle proteins, which are also active "kinetic" inhibitors of crystal growth/dissolution (62), but which certainly do not compromise the net beneficial effects of fluoride.

In support of this view of surface inhibitor reactivity for anionic anti-calculus active agents, preliminary measurements of fluoride and PPi uptake into artificial carious enamel from calculus control formulations showed PPi uptake mostly at the exterior regions of the lesions, with F distributed more uniformly into the carious enamel (63).

At present, we do not know the ideal proportions of inhibitor/F for maximum cariostatic effect. Inhibitors will exhibit variable effects on the de- and remineralization processes due to (i) different strengths of inhibition, (ii) transport properties within carious enamel, and (iii) other ancillary effects such as plaque activity. For example, calculus control formulations containing zinc chloride along with NaF exhibited decreased cariostatic activity compared to NaF controls in the in vitro pH cycling and animal caries models utilized in this report (64), as well as fluoride uptake protocols (65). In an in situ denture chip panel this same zinc chloride based formulation similarly exhibited directionally lower remineralization activity than NaF control and PPi/NaF dentifrices (66). Recent clinical studies (67) showed an 11.4 percent decrease in anticaries efficacy for a zinc chloride/NaF dentifrice versus a PPi/NaF dentifrice but the reported differences were not statistically significant (p=0.99). Zinc citrate, on the other hand, when formulated at 0.5 percent in monofluorophosphate (MFP) based calculus control formulations, exhibited equivalent anticaries efficacy to MFP controls in both the pH-cycling and rat caries models (64) corroborating recent clinical studies (68) on this system. Recently, antitartar benefits of 2 percent zinc citrate formulation were reported (69) which were considerably larger than the benefits observed in previous calculus clinicals of the 0.5 percent zinc citrate/MFP formulation (70). The effects of this increased dosage of zinc citrate on profile or clinical caries performance, however, is unknown.

The group of laboratory studies described above which show clear correlation with caries clinical experience demonstrate significant anticaries activity for PPi/NaF (1100 ppm F) dentifrice combinations throughout the dosage range from 3.3-5.0 percent PPi. The overall benefits of the PPi/NaF system appear to include the cooperative effects of pyrophosphate and fluoride as dissolution inhibitors as well as the enhancement of remineralization by fluoride. While there remains the possibility for improving cariostatic efficacy by virtue of inhibitor effects on enamel dissolution, a long sought goal in inhibitor research (71,72), this would require greater knowledge of the synergistic effects of various agents on de- and remineralization. Different inhibitor-fluoride systems appear to behave quite differently in various profile test systems and thus caution must be exercised in extrapolating results directly to the oral environment.

For the PPi/NaF system it appears that the caries efficacy of F is not compromised throughout the dosage range clinically evaluated for calculus control. The research results to date suggest that our collective fears about remineralization interference and therefore increased caries due to anticalculus formulations were somewhat overstated (73-75). Solidly based profile systems, that can be matched with clinical experience, must be utilized in assessing the overall impact on caries efficacy of calculus control systems. Testing protocols which incorporate significant demineralization must be utilized in order to obtain proper indications of the likely clinical cariostatic potential of these anticalculus products.

Acknowledgement
This work was partially supported by NIDR Grant DE03223. The expert technical assistance of Ms. N. Barrett in the pH cycling study is acknowledged with thanks.

REFERENCES

1. Mallatt, M.E., Beiswanger, B.B., Stookey, G.K., Swancar, J.R. and Hennon, D.K. (1985) J. Dent. Res., 64, 1159-1162.

2. Zacherl, W.A., Pfieffer, H.J. and Swancar, J.R. (1985) J. Am. Dent. Assoc., 110, 737-738.
3. Lobene, R.R. (1986) Clin. Prev. Dent., 8, 5-7.
4. Schiff, T.G. (1987) Clin. Prev. Dent., 9, 13-16.
5. Lobene, R.R. and Volpe, A.R. (1987) Compendium Cont. Educ. Dent., Suppl. 8, 272-274.
6. Lobene, R.R. (1987) Compendium Cont. Educ. Dent., 8, 175-178.
7. Juliano, G.F., Villanueva, E.H. and Carlos, J.M. (1986) J. Philip. Dent. Assoc., 36, 28-30.
8. Schiff, T.G. (1987) Compendium Cont. Educ. Dent., Suppl. 8, 275-277.
9. Rosling, B. and Lindhe, J. (1987) Compendium Cont. Educ. Dent., Suppl. 8, 278-282.
10. Lu, K.H., Ruhlman, C.D., Chung, K., Adams, A. and Bollmer, B. (1988) J. Indiana Dent. Assoc., 67, 17-18.
11. Rustogi, K.N., Volpe, A.R. and Petrone, M.E. (1988) Compendium Cont. Educ. Dent., 9, 78-79.
12. Fleisch, H. and Bisaz, S. (1962) Am. J. Physiol., 203, 671-675.
13. Fleisch, H., Russell, R.G.G., Bisaz, S., Termine, J.D. and Posner, A.S. (1968) Calc. Tiss. Res., 2, 49-59.
14. Jung, A., Bisaz, S. and Fleisch, H. (1973) Calc. Tiss. Res., 11, 269-280.
15. Francis, M.D. (1969) Calc. Tiss. Res., 3, 151-162.
16. Fleisch, H., Russell, R.G.G. and Straumann, F. (1966) Nature, 212, 901-903.
17. Draus, F.J., Lesniewski, M. and Miklos, F.L. (1970) Arch. Oral Biol., 15, 893-896.
18. Briner, W.W., Francis, M.D. and Widder, J.S. (1971) Calc. Tiss. Res., 7, 249-256.
19. Muhlemann, H.R. and Aeschbacher, M. (1970) Helv. Odont. Acta, 14, 30-31.
20. ten Cate, J.M., Jongebloed, W.L. and Arends, J. (1981) Caries Res., 15, 60-69.
21. Regolati, B. and Hotz, P. (1970) Helv. Odont. Acta, 14, 42-44.
22. Lu, K.H., Yen, D.J.C., Zacherl, W.A., Ruhlmann, C.D., Sturzenberger, O.P. and Lehnhoff, R.W. (1985) J. Dent. Child., 52, 449-451.
23. Triol, C.W., Volpe, A.R., Ripa, L.W. and Leske, G.S. (1988) J. Dent. Res., 67, 230 (No.941).
24. Mellberg, J.R., Castrovince, L.A., Rotsides, I.D., Hayes, J.C. and Deutchmann, M. (1987) Compendium Cont. Educ. Dent., Suppl 8, 267-271.
25. Featherstone, J.D.B., Shariati, M., Brugler, S., Fu, J. and White, D.J. (1988) Caries Res., in press.
26. White, D.J. and Faller, R.V. (1987) Caries Res., 20, 332-335.
27. Featherstone, J.D.B. (1983) in Demineralization and Remineralization of the Teeth, (ed. Leach, S., Edgar, W.M.) pp 89-110, IRL Press, Oxford.
28. ten Cate, J.M. (1984) in Cariology Today (ed. Guggenheim, B.) pp.231-236, Karger, Basel.
29. Arends, J., Nelson, D.G.A., Dijkman, A.G., and Jongebloed, W.L. (1984) in Cariology Today (ed. Guggenheim, B.) pp.145-158, Karger, Basel.
30. ten Cate, J.M. and Duijsters, P.P.E. (1982) Caries Res., 16, 201-210.
31. Featherstone, J.D.B., O'Reilly, M.M. and Shariati, M. (1986) in Factors Affecting De- and Remineralization of the Teeth (ed. Leach, S.A.) pp.23-34, IRL Press, Oxford.
32. O'Reilly, M.M. and Featherstone, J.D.B. (1987) Am. J. Orthop. and Dentofac. Orthop., 92, 34-40.
33. Purdell-Lewis, D.J., Groeneveld, A. and Arends, J. (1976) Caries Res., 10, 216-226.
34. Featherstone, J.D.B., ten Cate, J.M., Arends, J. and Shariati, M. (1983) Caries Res., 17, 385-391.
35. ten Cate, J.M., Shariati, M. and Featherstone, J.D.B. (1985) Caries Res., 19, 335-341.

36. White, D.J. and Featherstone J.D.B. (1988) Caries Res., 21, 502-512.
37. Standards for Fluoride Dentifrices, The Proprietary Association Subgroup on Fluoride Dentifrices, Submitted to the FDA Panel on OTC Dentifrices and Dental Care Agents, Method No. 37, March 11, 1978.
38. Keyes, P.H. (1958) J. Dent. Res., 37, 1088-1099.
39. Navia, J.N. (1977) Animal Models in Dental Research, pp.287-290, Univ. of Alabama Press, Univ. of Alabama.
40. White, D.J. (1987) Caries Res., 21, 228-242.
41. Chen, W.C. and Nancollas, G.H. (1986) J. Dent. Res., 65, 663-668.
42. Newbrun, E. (1978) Fluorides and Dental Caries, 181pp., C. Thomas, Springfield, IL.
43. Brown, W.E., Gregory, T.M. and Chow, L.C. (1977) Caries Res., 11 (Suppl. 1), 118-141.
44. Moreno, E.C., Kresak, M. and Zahradnik, R.T. (1977) Caries Res., 11, (Suppl. 1), 142-171.
45. Fleisch, H. (1981) J. Crystal Growth, 53, 120-134.
46. Williams, G. and Sallis, J.D. (1982) Calc. Tiss. Res., 34, 169-177.
47. White, D.J., Bowman, W.D. and Nancollas, G.H., this issue.
48. Lilenthal, B. (1977) Phosphates and Dental Caries (ed. Myers, H.M.) 107 pages, Karger, Basel.
49. Arends, J., Dijkman, A.G., Ruben, J. and Jongebloed, W.L. (1988), this issue.
50. Stephen, K.W., Russell, J.I., Creaner, S.L. and Burchell, C.K. (1987) Caries Res., 21, 162.
51. White, D.J. (1988) Am. J. Dent., in press.
52. White, D.J. and McClanahan, S.F. (1988) J. Dent. Res., 67A, 114 (No.16).
53. White, D.J. (1987) Caries Res., 21, 228-242.
54. Feagin, F.F., Clarkson, B.H. and Wefel, J.S. (1985) Caries Res., 19, 219-227.
55. Featherstone, J.D.B.. Duncan, J.F. and Cutress, T.W. (1979) Archs. Oral Biol., 24, 101-112.
56. Margolis, H.C., Moreno, E.C. and Murphy, B.J. (1986) J. Dent. Res., 65, 23-29.
57. Arends, J. and Christoffersen, J. (1986) J. Dent. Res., 65, 2-11.
58. ten Cate, J.M., Duijsters, P.P.E. (1983) Caries Res., 17, 513-519.
59. Featherstone, J.D.B. (1984) Cariology Today (ed. Guggenheim, B.) pp.259-268. Karger, Basel.
60. Nelson, D.G.A., Featherstone, J.D.B., Duncan, J.F. and Cutress, T.W. (1983) Caries Res., 17, 200-211.
61. Wong, L., Cutress, T.W. and Duncan J.F. (1987) J. Dent. Res., 66, 1735-1741.
62. Moreno, E.C., Varughese, K. and Hay, D.I. (1979) Calc. Tiss. Int., 28, 7-16.
63. White, D.J., Sammons. M.C., Deibel, R.M., Coombs, M.A. and Lueders, R.A. (1988) J. Dent. Res., 67A, 115(No.17).
64. Featherstone, J.D.B., Shariati, M. and Brugler, S. (1987) J. Dent. Res., 66A, 242 (No.1087).
65. White, D.J. and Faller, R.V. (1987) Caries Res., 21, 40-46.
66. Mellberg, J.R., Castrovince, L.A., Rotsides, I.D.. Hayes, J.C. and Deutschman, M. (1987) J. Dent. Res., 66A, 108 (No.14).
67. Triol, C.W., Volpe, A.R., Ripa, L.W. and Leske, G.S. (1988) J. Dent. Res., 67A, 230 (No. 941).

68. Stephen, K.W., Russell, J.I., Creanor, S.L. and Burchell, C.K. (1987) Caries Res., 21, 162.
69. Kazmierczak, M., Mather, M.L., Ciancio, S.G., Fishman, S. and Ciancio, L. (1988) J. Dent. Res., 67A, 246 (No. 1068).
70. Lobene, R.R. (1987) Clin. Prev. Dent., 9, 3-8.
71. Anbar, M., St. John, G.A. and Scott, B.C. (1974) J. Dent. Res., 53, 867-878.
72. Bartels, T., Schuthof, J. and Arends, J. (1979) J. Dent., 7, 221-229.
73. Woltgens, J.H.M. and Koulourides, T. (1983) Caries Res., 17, 357-364.
74. Regolati, B. and Muhlemann, H.R. (1970) Helv. Odont. Acta, 14, 37-42.
75. Koch, G., Karlsson, R., Bergman-Arnadottir, I., Finnbogason, S., Hoskuldsson, O. and Bjarnason, S. (1988), this issue.

SESSION V.
Clinical Aspects of
Calculus (vs Caries)

A.Thylstrup
L.Chironga
J.de Carvalho
K.R.Ekstrand

The occurrence of dental calculus in occlusal fissures as an indication of caries activity

Department of Cariology and Endodontics,
Royal Dental College, Nørre Allé 20,
DK-2200 Copenhagen N, Denmark

Abstract

This study was undertaken to examine the relation between occlusal attrition, caries and indications of calculus in fissures of posterior teeth. In order to study the natural interplay between usage of teeth and occlusal caries, it was necessary to examine extracted teeth from a country with limited dental resources. The material comprised a total of 381 teeth extracted in Zimbabwe. 120 permanent molars were grouped into two categories: (1) 48 teeth with increasing surface wear and sparse bacterial deposits. (2) 72 teeth with little wear and different degrees of undisturbed bacterial deposits and carious destructions. The teeth were serially sectioned in bucco-lingual direction and the sections examined in a stereo-microscope and in polarized light. Teeth from (1) showed different stages of arrested occlusal caries irrespective of fissure depth. The fissures of the arrested lesions were most often completely filled with calculus. The combined clinical and microscopical examination of teeth from category (2) indicated that occlusal enamel breakdown was always related to the deepest or the most protected parts of the surface profile. The spread of enamel breakdown appeared to be a result of further demineralization from an initially established focus rather than a general demineralization involving the entire fissure system. Thus examinations of sections in the vicinity of the visual enamel destruction showed much less advanced stages of caries or even arrested lesions with calcified material occupying the narrow fissure. On the basis of these principal observations it is reasonable to expect that external control of dental plaque on occlusal surfaces promotes calcification of plaque in deep or inaccessible narrow fissures.

Recent Advances in the Study of
Dental Calculus

Introduction

The occlusal surfaces of posterior teeth have repeatedly been noted to be the most vulnerable locations for dental caries (1). Since the high incidence of caries on occlusal surfaces has for a long time been recognized by clinicians and scientists, it is no wonder that the unpredictable and narrow fissures of the occlusal surfaces have commonly been considered to be the focus of caries initiation on these surfaces (1-6).

Common clinical experience with occlusal caries was early in this century translated by restorative dentistry to invasive treatment techniques in terms of preventive fillings (1).

Thus Hyatt advocated the placement of small amalgam restorations in newly erupted teeth before the appearance of clinical signs of caries because "The enamel defect of today is the carious cavity of tomorrow" (7). Even though Hyatt's statement has lost its validity long ago it is still common to teach students during their first days at the dental school to place a perfect class I amalgam restoration in a completely sound and innocent plastic tooth. There is no doubt that the widely accepted and practiced technical management of the "vulnerable" occlusal surface without a proper understanding of the pathogenesis and pathology of caries on these surfaces has contributed to the formation of clinical as well as scientific myths on diagnosis and treatment of caries in fissures. The fissure sealant technique, where dental materials are used to "seal" caries-susceptible areas from the oral environment is just a modern version of the classical concept that technical procedures are required to prevent caries on occlusal surfaces. The potential toxic side-effects of dental materials, however, are currently receiving increasing public as well as professional attention in many countries (8). For this reason there is a growing demand for development of appropriate non-operative treatment techniques of occlusal caries. Diagnosis and particularly prognosis of occlusal caries require a profound understanding of the natural history of caries on these surfaces. Due to the widespread use of preventive filling techniques in industrialized societies it is necessary to initiate such studies using extracted teeth from a country with limited dental man-power resources.

The purpose of this study was to examine the possible relation between occlusal attrition, caries and light microscopical indications of calculus in posterior teeth extracted in Zimbabwe.

Material and methods

The material comprised a total of 381 permanent teeth extracted February - March 1985 in Harare, Zimbabwe. 138 permanent molars were grouped into two categories; (1) 48 teeth with different degrees of surface wear; (2) 72 teeth with no or only modest signs of occlusal wear and with different degrees of caries destruction. 18 molars were omitted from the study because of complete or almost complete destruction of the crown.

The teeth were stored in formalin and photographed in a stereo-microscope before and after postfixation in osmic acid. The teeth were serially sectioned in bucco-lingual direction.

The mean number of sections per specimen was 11 and the thickness of each section was 200 μm. Before separating the sections the teeth were photographed in a stereo-microscope. In this way it was always possible to relate the histological observations to the macro-morphology of the occlusal surface. The sections were examined and photographed in a stereo-microscope. In addition to this some of the sections were selected and prepared for examination in polarized light.

Results

Group I
The teeth in this group were classified according to 4 grades of occlusal wear.

1.   General wear of cusps and initial levelling out of ridges.
2.   Marked cuspal wear with a levelling out of ridges and grooves.
3.   Local exposure of dentine.
4.   Complete levelling out of the occlusal surface profile with different degrees of dentinal exposure; remnants of grooves and fissures can still be discerned.

The clinical and stereo-microscopic examination revealed a spectrum of signs of caries at the proximal surfaces. Few cases (Fig. 2) had isolated progressive caries with exposure of dentine. The most common observation was marked proximal wear combined with arrested caries in different stages of progression. Figs. 1 to 4 indicate progressive degrees of wear corresponding to the classification. Examination in the stereo-microscope prior to postfixation with osmic acid showed that all teeth independant of degree of wear had dark stained stria corresponding to the remaining parts of the occlusal grooves and fissures. The postfixation with osmic acid which stains organic material and bacterial deposits revealed that none of the teeth with signs of marked wear were entirely free of dental plaque. However, with increasing occlusal wear the bacterial deposits appeared to be increasingly concentrated corresponding to remnants of the grooves and fissures. None of the sections examined in this group of teeth showed light microscopic evidence of progressing caries in relation to the grooves and fissures. It was of particular interest that no "undermining" caries was observed even where the bottom of deep and narrow fissures were near the dentinal surface (Figs. 1a,b; 2a,b; 3a, 4a). In contrast the majority of the sections, particularly those representing wear of 3rd and 4th grade, were filled with calcified material occupying the narrow fissure (Figs. 3a,b; 4a,b). In spite of the grinding procedure needed for examination of sections in polarized light it was possible to identify remnants of dental plaque corresponding to those observed prior to the sectioning (Figs. 3b, 4b).

Group II

The teeth in this group were classified into 4 types of progressive destructions.

1.   Minor destructions of the original surface morphology.
2.   Well-defined destruction confined to the enamel.
3.   Well-defined enamel destruction with possible dentinal exposure.
4.   Marked cavitation with dentine destruction.

The clinical and stereo-microscopical examination of these teeth disclosed a wide spectrum of caries on proximal as well as on lingual and buccal surfaces. The majority of lesions were characterized as active. The indications of interproximal wear were less pronounced than in group I. Figs. 5 to 8 show progressive stages of caries destruction according to the classification system. It was a characteristic feature that early but clinically visible destruction always appeared as a localized phenomenon and that the localization of the destruction within the same type of molars corresponded to specific areas of the macro-morphology. Compare for example Fig. 5 with Fig. 6.

Consequently, we never clinically observed a diffuse or general attack involving the entire fissure system. The postfixation with osmic acid revealed not surprisingly larger accumulations of plaque than in group I teeth (Figs. 5a to 8a). Regardless of amount of plaque the most prominating cuspal parts were free of visible plaque. Even though considerable parts of the occlusal surface were covered with plaque, visible destructions were only seen in relation to the thickest and most stained plaque accumulations.

The light microscopical examination of sections cut through the local destruction and sections representing the adjacent groove or fissure part without visible destruction confirmed the clinical observation that caries destruction on the occlusal surface is the result of a local progressive process. Thus Figs. 5b to 8b illustrate increasing degrees of hard tissue destruction. Fig. 5b shows a small cavity developed in a groove. The cavity is sufficiently large to harbour stained plaque. The adjacent fissure had no signs of progressive destruction (Fig. 5b). Fig. 6b shows a more extensive cavity with the same anatomical localization as the one seen in Fig. 5. The cavity involves the outer third of the enamel whereas the underlying enamel is opaque indicating modest loss of mineral. Two separated areas along the enamel-dentinal border appears dark brownish indicating superficial dissolution of the dentine. These two areas correspond to two particularly active processes in the cavity bottom. Due to the corresponding difference in the orientation of the underlying enamel rods the initial dentinal dissolution appears as two seperated reactions.

Fig. 7b shows a characteristic, more advanced stage of cavitation with the dentine being directly exposed to the bacteria in the cavity. The shape of the cavity corresponds to the direction of the rods, and the dentine is not involved beyond the area of demineralized enamel. The adjacent fissure and groove show no signs of caries (Fig. 7c). The narrow fissure is filled with calcified material. The size of the local destruction

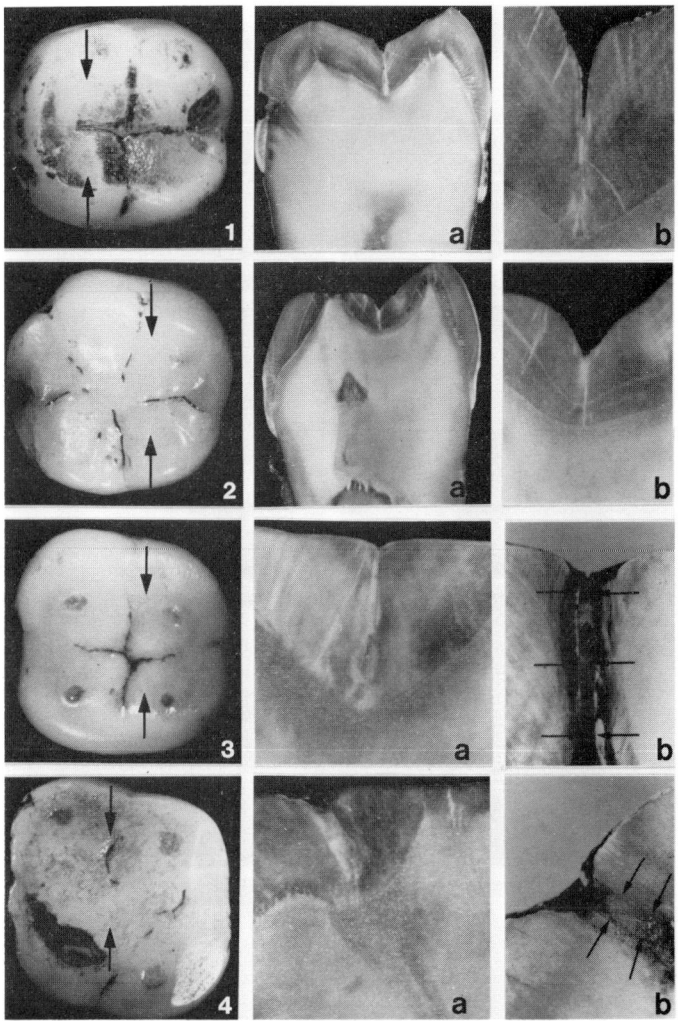

Figs. 1-4. Examples of teeth with progressive signs of occlusal wear according to the classification. The teeth are photographed after staining of organic material with osmic acid. Note an active lesion with cavitation at the distal surface in Fig. 2. Fig. 3 shows an arrested lesion in conjunction with marked interproximal wear at the mesial surface. Figs. 1a and 2a. The sections are localized corresponding to arrows in Figs. 1 and 2. Note signs of caries in the buccal pit in Fig. 1a. Figs. 1b and 2b. Details of fissure area with remnants of calcified material. Figs. 3a and 4a. Details of fissure areas corresponding to arrows in Figs. 3 and 4. Figs. 3b and 4b. Details of fissures photographed in polarized light after imbibition in quinoline. Arrows mark the walls of the fissures. The material occupying the fissures is calcified.

A.Thylstrup *et al.*

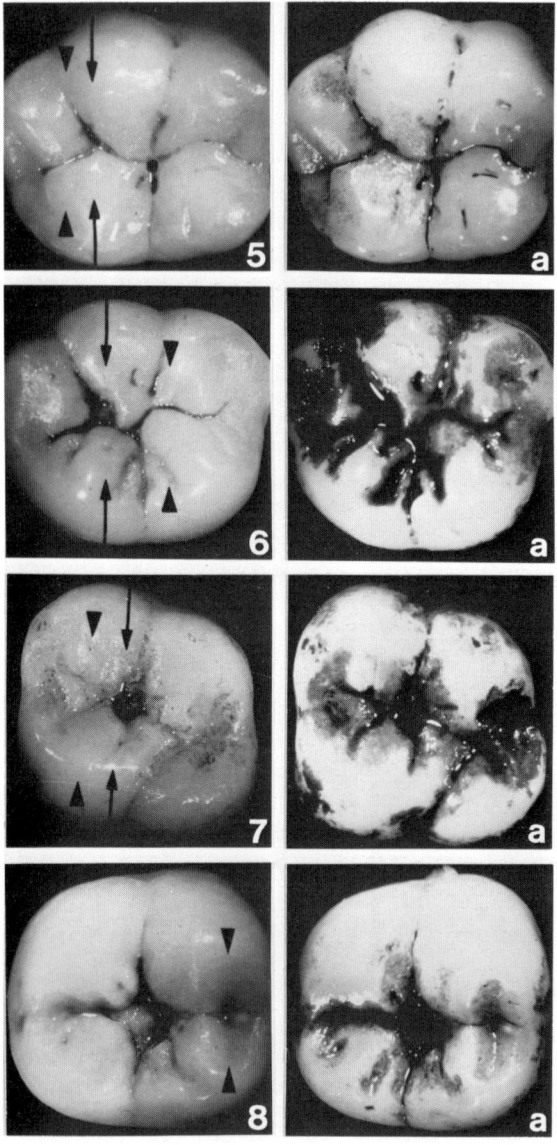

Figs. 5 - 8. Progressive stages of enamel and dentine destruction according to the classification. Figs. 5a to 8a show the
same teeth after staining of organic material with osmic acid.
Figs. 5b to 7b. Sections corresponding to localized destructions
marked with arrows in Figs. 5 to 7. In Fig. 5b caries breakdown
is seen in the groove whereas the adjacent fissure (F) shows no
signs of active demineralization. D indicate demineralization
and discoloration of dentine (Figs. 6b and 7b). Note that

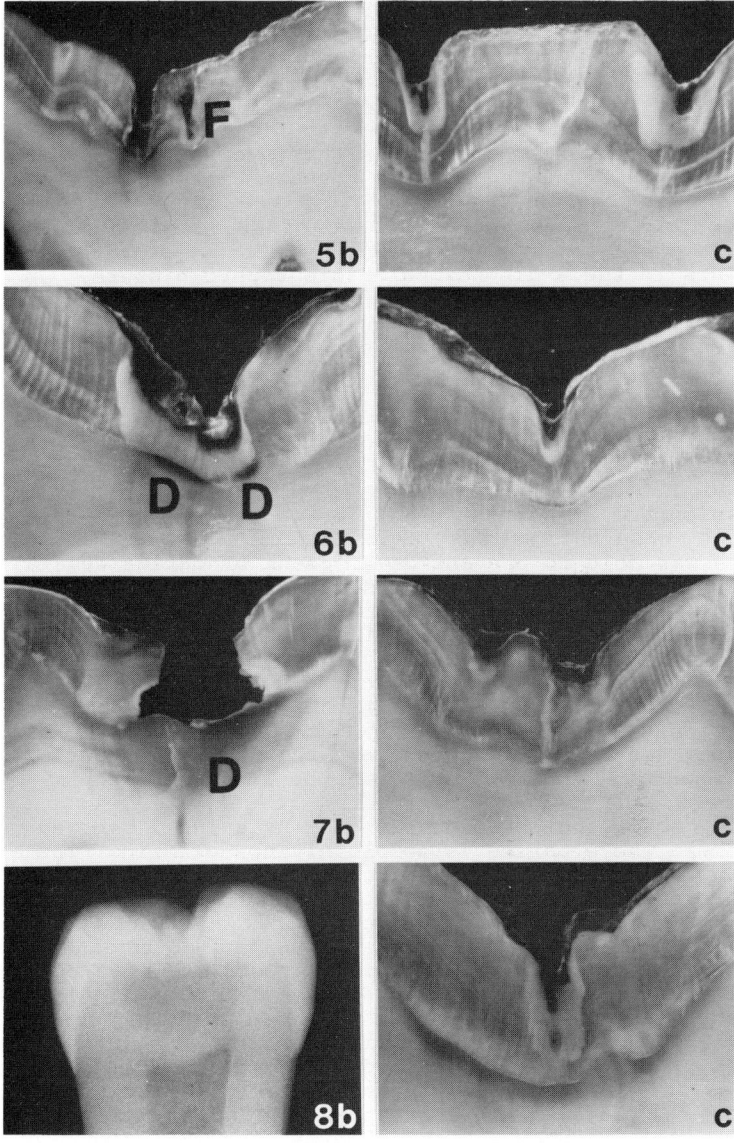

demineralization of the dentine is not spreading along the enamel-dentinal junction but occurs only corresponding to the involved enamel. Fig. 8b. Radiograph of specimen in Fig. 8. Figs. 5c to 8c. Sections of grooves and fissures in the vicinity of the local destructions seen in Figs. 5 to 8 (triangles). Irrespective of degree of local destruction the adjacent grooves and fissures show only modest signs of demineralization. The fissure in Fig. 7c is filled with calcified material.

shown in Fig 8 is illustrated by the use of a radiograph (Fig. 8b), as this underlines the classical clinical impression of the undermining nature of pit and fissure caries. On the radiograph the occlusal surface appears as almost "intact" in contrast to the extensive dentinal destruction. The clinical picture in Fig. 8a, however, clearly shows the cavity filled with dental plaque. Due to the orientation of rods in the involved enamel the cavity is shaped as a truncated cone in the same manner as seen in Fig. 7b. For this reason the opening of the cavity is relatively small compared to the large amount of surrounding sound enamel which again means that the clinically distinct opening apparently disappears in the 2-dimensional radiographic picture.

The histological changes corresponding to the adjacent grooves and fissures (Figs. 5c to 8c) showed thin plaque accumulations and modest enamel demineralizations.

Discussion

The classification and examination of occlusal surfaces of teeth extracted in a country with limited dental man-power resources revealed important principles with regard to initiation, progression and arrestment of caries on these surfaces. The most important observation was that functional wear on occlusal surfaces was able to prevent development of caries regardless of the depth of remaining narrow fissures.

It is tempting to interpret these clinically "sound" surfaces as the result of a non-cariogenic environment in its broadest sense. This means that the individual tooth as well as the individual person have been born and living in a society where the cariogenic challenge is low due to, for example, a low consumption of refined carbohydrates. Closer examination of the individual teeth revealed, however, that the other surfaces with less pronounced wear had different degrees of signs of past or present caries activity. For this reason it is possible to conclude that the examined teeth have not emerged in an environment where caries is completely non-existing. The absence of signs of caries on the worn occlusal surfaces in group I therefore indicates that continuous disturbance and wearing away of microbial accumulations on the surface itself are able to prevent caries progression in occlusal grooves and fissures irrespective of the depth of these structures. This observation has important clinical implications as it clearly indicates that efficient control of plaque on the occlusal surface prevents lesion initiation and progression even in narrow and deep fissures. The conclusion accords with findings demonstrating that lesions in narrow fissures are initiated at the fissure entrance (9, 10), which corresponds to areas where growth of bacteria has continuously been prevented or disturbed on teeth with signs of heavy wear.

The classification of teeth without obvious signs of occlusal wear and hence more extensive plaque accumulations made it possible to examine progressive stages of carious destruction on occlusal surfaces.

The fact that lesion formation on occlusal surfaces seemed to be initiated by a local destruction was of particular interest. The localization of the initial signs of carious destruction appeared also closely related to the tooth specific macro-morphology. Due to the limited number of available teeth it was not possible to precisely identify the localization of the first signs of destruction with regard to the different types of permanent molars. The general impression was, however, that initial destruction was located corresponding to the deeper parts of the macroscopic groove-fossa-system. In general terms the initial destruction seems to be located within the occlusal surface at sites where bacterial accumulations receive the rela-tively best protection against functional wear. The clinical as well as the microscopical examination of teeth with larger cavi-ties confirmed that the visible destruction was a local phenome-non as adjacent grooves and fissures had no or only modest signs of demineralization. This observation was independant of the size of the local destruction. It is therefore tempting to conclude that progressive destruction of occlusal surfaces is most often initiated by a local process in the deepest part of the groove-fossa-system due to accumulation of undisturbed bacterial deposits. In this area, which already offers a relative protection to micro-organisms against physical wear, the forma-tion of a cavity further improves local conditions for lodgement and growth of oral bacteria. This accelerates demineralization and destruction, which again improves local conditions for growth of bacteria.

In order to understand the rapid progression of occlusal caries under natural conditions it is necessary to appreciate the particular anatomical configuration of the occlusal surface where caries is initiated. Firstly it is important to understand the process in 3 dimensions as caries on occlusal surfaces is most often initiated in depressions where two or more grooves meet (11). Such depressions are defined as "fossae" in the descriptive system of the fundamental macro-morphology of teeth proposed by Carlsen (12). This means that several "surfaces" are involved in the initial enamel dissolution on occlusal surfaces. Because enamel demineralization always follows the direction of the rods (13), it is natural that the enamel lesion initiated in a fossa gradually assumes the shape of a cone with its basis towards the enamel-dentinal junction. The dentine reaction to progressing caries corresponds to the direction of the involved enamel rods (13). For this reason sections cut through a such lesion gives the 2-dimensional impression of two seperated lesions (Fig. 6b). In a fossa, however, where several "surfaces" are involved the lesion entity is in reality shaped as a cone in three dimensions. Thus it is no wonder that clini-cians and textbooks on operative dentistry often pay special attention to the socalled "undermining" character of occlusal caries. However, in light of the structural arrangement of the enamel rods in the occlusal groove-fossa-system the mode of lesion growth in these areas is not particularly surprising. Furthermore, in accordance with previous observations on proximal lesions (13) we did not in this study notice dentine reactions beyond the border, where the rods involved in the enamel lesion reached the dentine (Figs. 6b, 7b). With progressing enamel destruction a proper enamel cavity is formed (Fig. 7b). Again the outlines of the cavity reflect the arrangement of enamel rods in the area. The cavity is thus shaped as a truncated cone.

The particular anatomical configuration of the part of the occlusal surface where caries begins explains why the openings of occlusal cavities is always smaller than the basis. The "closed" nature of the process obviously favours undisturbed growth of oral bacteria and hence further destruction of the tissue. The "closed" nature of the process is therefore responsible for the extensive dentinal destruction frequently observed with only modest enamel destruction due to the heavy accumulations of bacteria harboured in the protected environment. (Figs. 8a, b). On the basis of our observations of the natural history of occlusal caries, it is possible to conclude that occlusal enamel breakdown probably most often commences in the deepest or most protected parts of the external surface profile. The spread of enamel breakdown is thus a result of further demineralization from an initially established focus rather than a general demineralization involving the entire fissure system. A similar conclusion was drawn by König (9) who also related observations on sections to the macro-morphology of the occlusal surfaces of permanent premolars.

However, the major part of clinical and scientific concern with regard to occlusal caries has been devoted to the possible events taking place in deep and inaccessible fissures (1). For this reason, it is important to realize that caries destruction is always initiated on the surface of enamel due to metabolic activities in bacterial accumulations on the surface. It is reasonable to assume that evolution of plaque with cariogenic potential requires space which in this context is available only above the entrance to narrow fissures (9). This assumption is supported by the few studies on the ultrastructural features of the organic content in narrow fissures. Thus Theilade et al. (14) observed different degrees of mineralization resembling calculus in occlusal fissures only seven days after exposure to oral bacteria. This observation was interpreted as an indication of non-cariogenic bacterial accumulations particularly in the deepest part of the fissure. Similar but more extensive formations of calculus have been described in fissures of teeth which had been in the oral environment for prolonged periods (15, 16). In our study we observed that narrow fissures in teeth with signs of marked attrition were almost completely filled with calculus. The absence of any signs of active carious dissolution therefore makes is reasonable to believe that the presence of calculus in fissures is a strong indication of bacterial accumulations with no or extremely little cariogenic potential.

In a general perspective it is clear that a gradual opening up of a narrow fissure approaching the V-shape of a groove provides more favourable conditions for growth of bacteria than those in the narrow fissure. For this reason conditions for bacterial growth in grooves can be compared with those at the entrance to the fissure proper (14). This assumption was recently supported by light microscopic and ultrastructural examinations of characteristic fissures and grooves in erupting molars (17). Thus bacteria in deep fissures were avital constrasting viable bacteria in grooves and at the entrance to the fissures. This study also demonstrated the relationship between the growth conditions for the bacteria and their cariogenic potential as caries was only observed in grooves and at the fissure entrance.

The clear relationship demonstrated in this study between wear and caries progression gives reason to propose that the time between eruption and full occlusion is the most critical period for initiation of occlusal caries because of reduced functional usage and wear of erupting teeth. Considering that we are in this study dealing with teeth from a country with limited dental manpower resources, it is tempting to hypothesize that the advanced lesion in Fig. 8 is the result of a minor local destruction (Fig. 5) established during eruption. Examination of teeth in different stages of eruption at the ulstrastructural level indicate significant changes in plaque distribution and enamel reactions which could be related to changes in degree of participation of the teeth in mastication (18). This observation was recently confirmed at the clinical level, as partially erupted first permanent molars had significally higher plaque scores on the occlusal surfaces than fully erupted molars (11). Correspondingly a higher proportion of incipient active enamel lesions was observed on the occlusal surfaces of erupting molars than on fully erupted teeth.

On the basis of numerous epidemiological data it is widely accepted that the onset of caries is closely related to children in general and more specifically to newly erupted teeth. Even though the effect of a socalled postmaturation process of newly erupted teeth has never been proven, it is nevertheless common to explain the particular susceptibility of young teeth to dental caries as a result of their immature status. Erupting teeth are not particularly immature, however, but as they do not participate in mastication they offer more favourable conditions for growth of oral bacteria than fully erupted teeth. So far little interest has been paid to the importance of usage of teeth and wear for the maintenance of tooth integrity in the oral environment (18). In this context it is interesting briefly to discuss the observations of Davies and Pedersen (19) on occlusal attrition in East and West Greenland. The study was carried out in 1937 in order to examine the influence of urbanization on occlusal wear. They recorded a dramatic difference in degree of attrition as even 10-14-year-old children in East Greenland had markedly worn teeth. The similar degree of wear did not occur in West Greenland until 30-39 years of age and even then 35% of the teeth were still unworn. The authors also noted that among a total of 13,308 permanent teeth examined in East Greenland only 68 surfaces were carious. 50 of the carious surfaces were occlusal and of these were 30 in "largely unworn" third molars. In the same period of time a dramatic increase in caries prevalence was noted in West Greenland (20). It has been widely accepted to see this increase as a result of altered living conditions with particular emphasis on increased consumption of refined carbohydrates (21) in accordance with the prevailing concept of the cariogenecity of sucrose (1).

The marked change in usage and hence wear of teeth noted in West Greenland in Davies and Pedersen's pertinent study gives, however, reasons to reconsider the simple relationship commonly held between the rise in caries and increased consumption of high-sugar diet in Greenland. The first step in this direction may be to understand the relationship between diet and caries in a broad physiologic perspective rather than being a question only about the amount of refined carbohydrates.

A.Thylstrup *et al.*

References

1. Newbrun, E. (1983), Cariology, 2nd ed., Williams and Wilkens, Baltimore, London.
2. Bödecker, C.F. (1928), Dent. Items Interest, 50, 7-11.
3. Parfitt, G.J. (1955), Brit. Dent. J., 99, 423-427.
4. Goose, H.D. (1972), Publ. Health, 86, 137-141.
5. Bergman, D.S., Slack, G.L. (1973), Brit. Dent. J., 134, 135-139.
6. Thomsen, J.R., Tagesen, J., Fejerskov, O. (1988), Tandlægebladet, 92, 1-6.
7. Hyatt, T.P. (1933), Prophylactic Odontotomy, MacMillan, New York.
8. Prevention and treatment of dental caries. Consensus report (1986), Dan. Med. Bull., 33, 199-202.
9. König, K.G. (1963), J. Dent. Res., 42, 461-476.
10. Mortimer, K.V. (1964), Adv. Fluor. Res., 2, 85-94.
11. Carvalho, J.C., Ekstrand, K.R., Thylstrup, A. (1989), J. Dent. Res., in press.
12. Carlsen, O. (1987), Dental Morphology, Munksgaard, Copenhagen.
13. Thylstrup, A., Qvist, V. (1987), in Dentine and Dentine Reactions in the Oral Cavity (eds. Thylstrup, A., Leach, S.A., Qvist, V.) pp.3-16. IRL Press Ltd., Oxford.
14. Theilade, J., Fejerskov, O., Hørsted, M. (1976), Archs. oral Biol., 21, 587-598.
15. Frank, R.M. (1973), Archs. oral Biol., 18, 9-25.
16. Galil, K.A., Gwinnet, A.J. (1975), Archs. oral Biol., 20, 559-562.
17. Ekstrand, K.R., Westergaard, J., Thylstrup, A. (1988), Caries Res., 22, 107.
18. Holmen, L., Thylstrup, A. (1986) in Tooth Morphogenesis and Differentiaiton II (eds. Belcourt, A.B., Ruch, J.-V.) pp. 283-294. INSERM vol. 125, Paris.
19. Davies, T.G.H., Pedersen, P.O. (1955), Brit. Dent. J., 99, 35-43.
20. Pedersen, P.O. (1966), Odontol. Revy, 17, 99-100.
21. Møller, I.J., Poulsen, S., Nielsen, V.O. (1972), Scand. J. Dent. Res., 80, 169-180.

F.Manji[1]
O.Fejerskov[2]
V.Baelum[2]
N.Nagelkerke[1]

Dental calculus and caries experience in 15 – 65-year-olds with no access to dental care

[1]Kenya Medical Research Institute, Medical Research Centre, Nairobi, Kenya and [2]Royal Dental College, Aarhus, Denmark

ABSTRACT

Although the conditions favouring mineral deposition and those causing caries dissolution are mutually exclusive, clinical data from epidemiological studies in children and adults studying the association between these two conditions are inconclusive. In the present report the relationship between supragingival calculus and dental caries has been studied in 1131 adults aged 15 to 65 years living in a rural area of Kenya. The population is having bad oral hygiene and limited access to dental care. In general a weak negative association between calculus and caries experience was found although in a small proportion of individuals the presence of the two conditions on the same surface was found to occur more often than was expected. The findings emphasize the complex physiological and physico-chemical conditions prevailing which contribute to the deposition of calculus.

INTRODUCTION

Dental caries develops only underneath the microbial deposits
on the teeth. Likewise dental calculus is not formed on the
tooth surface if there are no miccroorganisms present. Well
known and simple statements, but very important to have in
mind when discussing the relationship between dental calculus
formation and dental caries experience.

Although saliva is supersaturated with respect to hydroxyapatite (1,2,3), calculus does not form in all individuals
harbouring dental plaque, and it would be expected that patients forming large amounts of calculus - all other variables
kept equal - might be less susceptible to caries. Theoretical-
ly, the conditions favouring mineral deposition and those causing caries dissolution are mutually exclusive.

However, few experiments have been conducted under sufficiently controlled conditions in vivo to investigate the extent
to which calcium and phosphate deposits in plaque can influence initiation and progression of dental caries (4,5). Similarly only limited epidemiologic data on the association between
dental calculus and caries have been published. Direct comparison of the findings of various reports is hampered both by
the very different diagnostic criteria used and the different
ways, in which data have been analysed.

In children, Stones et al. (6) reported on results obtained
from a study (in 1942) of 280 children aged 3 to 16 years living in a National Children's Home. Caries was recorded as 0 =
tooth sound; 1 = one or more cavities in the same tooth using
probe or X-ray), the lesion confined to enamel, only; 2 = dentin involved but less than a quarter of the crown destroyed; 3
= more than a quarter of the tooth crown destroyed. Calculus
was recorded for each tooth as either 0 = no calculus; 1 =
least detectable amount; and 3 = heavy deposits. The relationship between caries and calculus in permanent teeth was studied by averaging calculus scores for the lower canines and incisors and relating these scores to the average caries scores
for the first molars in upper and lower jaws. Overall their
data showed little, if any, association between caries in the
molars and the amount of calculus on lower anterior teeth.

In contrast, James (7) reported on the prevalence of calculus in relation to mean DMFT values in 589, 11-12 year-old
children from a low fluoride area. The children were classified as having or not having calculus, calculus being situated
mostly subgingivally behind the lower incisors. Calculus was
observed in 38% of the children, and such individuals had a
mean DMFT of 4.6-4.8, as compared with a mean DMFT of 5.6-6.2
in those without calculus, demonstrating an expected slight
inverse relationship between calculus and caries experience.
Such findings were supported by Marthaler and Schroeder (8)
who studied the relationship between supragingival calculus on
lower incisor teeth and DMF experience in a total of 4317
children aged 8-15 years. Caries was recorded on particularly
susceptible surfaces of the right side of the mouth (the total
number of sites being 46). On an average children with calculus showed a slightly lower DMF experience in lower and upper
incisors (in smooth surfaces and in pits and fissures) as com-

pared with children without calculus. Only because of the large sample size were the small differences between groups statistically significant.

Sutfin et al. (9) studied 164 children aged 6 to 13 years in two different Guatemalan villages and found that children exhibiting calculus (or what was called "calcified plaque" which had a gritty consistency and was considered a precursor of fully calcified calculus) on buccal surfaces of mandibular first molars and lingual surfaces of maxillary first molars had much less caries on these two sites than children without such deposits. In their analyses, however, each surface was treated as if it was independent of the other sites in the same individual, and thus interpretation of the data is difficult. However, 38% of the children in one village had caries (on these teeth) with 81% exhibiting "calcified plaque" (17.7% with gross calculus), whereas in the other village 55% had caries and 40% "calcified plaque" (2.4% with gross calculus). The differences between the villages were attributed to differences in diet, although each village also represented individuals of quite different socio-economic status.

In adults, available data are scanty and inconclusive. Jacobson and Kesel (10) selected 102 patients in the age between 25 to 60 years from a dental school and classified them into three groups: calculus-free (40), calculus-present (40) and periodontal-diseased (22). The calculus-present group was further subdivided into "slight" (13 patients) with small amounts of calculus deposits (usually lingual surfaces of lower incisors and buccal surfaces of upper molars); "moderate" (25 patients) characterized by larger amounts usually on the same sites coming to a thickness of the width of the marginal gingiva, but other teeth could also be involved; "abundant" (2 patients) having excessive amounts of deposits on many of the teeth. Only unfilled, open carious lesions were recorded. An inverse relationship between caries and calculus was found: 62-64% had no caries lesions in the calculus-present group in contrast to 32% in the calculus-free group. 15-20% had 1-4 carious lesions in the slight and moderate calculus groups whereas 47.5% were thus classified in the patients with no calculus. However, in the group of individuals with > 5 carious lesions no obvious differences were found between slight, moderate and no calculus individuals (23.2%, 16% and 20%, respectively).

Opposing these observations, Little et al. (11) failed to find any relationship between the amount of dental calculus deposition and caries activity in calculus producers. Out of 1,144 children and adults attending the Eastman Dental Clinic in Rochester, N.Y., 295 were considered calculus producers. Calculus (both supra- and subgingival) was recorded according to Davies (12). Patients were at the same time classified as to their caries activity as follows: "active caries" - in which all carious lesions were soft and spongy, or penetrating into dentin as seen on bite-wing radiographs; "inactive caries" - all lesions were hard and glossy or did not penetrate beyond the E-D junction; "active-inactive caries" - both conditions present; "restored caries" - no evidence of ongoing disease but fillings present; "caries-free" - neither caries

nor fillings. About 70-90% of every age group had caries, but
after 40 years of age there was a significant decrease in the
proportion of subjects having "active caries" (from 60-80% to
30-45%).

To test whether conditions classified as "active caries"
would be found along with calculus, 42 of the heaviest calculus producers were placed on 4-months recall following removal of all calculus. There was no change in caries experience with increasing deposition of either supra- or subgingival calculus. The caries distribution was similar irrespective of whether the deposits were slight or heavy. Of those
with supragingival calculus, 38.5% had calculus deposits on at
least one carious tooth, and an average of 87% of these lesions were classified as "active". In addition, 69% of the
heavy calculus producers had "active" caries and on recall
most of these lesions were considered to have progressed, as
shown by X-ray, and new calculus had also been deposited.

In a study on 154, 19-22 year-old Finnish men, Ainamo (13)
studied the relationship between caries experience recorded as
either "healthy", "decayed", "filled" or "missing" and calculus measured according to the criteria used in the so-called
retention index by Björby and Löe (14). Separate index values
of retentive calculus (RC) were recorded. The author confirmed
the well known clinical observation that lower first molars
are more often decayed and extracted than the upper, whereas
incisors in the lower jaw showed abundant calculus, but decay
occurred more often in the upper anterior teeth. On a tooth
surface level, abundant supragingival calculus was found to
"prevent dental decay" whereas caries lesions were found more
often on tooth surfaces covered by subgingival calculus. Ainamo's data showed that the correlation coefficient between DMFS
and RC was -0.178 (p < 0.05), indicating that the presence of
calculus explained only about 3-4% of the variation in caries
experience. At tooth level, correlation coefficients between
retentive decay or retentive fillings and calculus were -0.10
and -0.11, respectively. Finally on surface level the values
were -0.04 and -0.06.

Berkey et al. (15) analyzed data obtained from the Veterans
Administration Dental Longitudinal Study (16) including 602
individuals who at the start were 28 to 76 years old and who
had at least 10 teeth present. Caries was recorded both clinically and radiographically, while calculus was classified for
each tooth as rare, discontinuous flecks, "non-continuous or
on one tooth surface", and "circumferential band around
tooth". During the 10-year follow-up 1078 enamel lesions on
proximal surfaces in 412 persons were recorded, but those surfaces originally recorded as sound but which had been subsequently restored were excluded from the analysis. Calculus was
found to be negatively associated with the number of filled
and diseased surfaces and with the number of teeth present.
Likewise, the level of calculus was significantly associated
with enamel lesion progression, but changes in calculus level
were not so associated.

The somewhat inconclusive nature of the findings of many
previous studies may have, in part, been due to the fact that
they have been performed in populations with considerable ex-

posure to dental care. We therefore found it of interest to investigate the relationship between supragingival calculus and dental caries in a free-living adult population living in a rural area with limited access to dental care.

MATERIALS AND METHODS

The data for this study originate from a survey of 1131 adults aged 15 to 65 years conducted in the Northern Division of Machakos District in Kenya. A complex random cluster sampling method was used to select individuals for examination (details of sampling procedures have been described elsewhere (17)). Each individual was examined for the status of the dentition, dental caries, mobility of the teeth, plaque, calculus, pocket depth, attachment level, and gingival bleeding. The findings in relation to caries and periodontal conditions have been described elsewhere (18,19).

Details of the diagnostic criteria for caries have been described elsewhere (18). For the purpose of the present analysis, caries has been defined as the presence of a lesion in which dentinal involvement was definitely suspected (D3), or where there was involvement of the pulp (D4), or where surface had a root-surface lesion (D8 or D9). Filled surfaces and missing teeth were excluded from all computations of caries experience, as were recordings of enamel lesions.

Calculus was scored on four surfaces of each tooth (mesial, facial, distal and lingual) as follows: 0 = none; 1 = supragingival calculus only; 2 = subgingival calculus; and 3 = supra- and subgingival calculus. For the following analysis, the presence of supragingival calculus was taken as the presence of either calculus score 1 or score 3.

RESULTS

The analyses here were performed around four null hypotheses:

Hypothesis 1
That there is no association between the occurrence of supragingival calculus on a given surface of a tooth and the occurrence of caries on the same surface. Fisher's exact test was used to test the hypothesis using each surface as the unit of analysis. This analysis was performed to assess the extent of the association across individuals for each single surface.

No association between caries and supragingival calculus was apparent for any surface, in any age group. In all cases, however, there was a tendency for the observed number of surfaces having both caries and calculus present on the same surface to be lower than the number expected (i.e. a weak tendency for a negative association). Table 1 provides, as an example, the observed and expected frequencies for facial surfaces exhibiting both calculus and caries (here presented irrespective of age group).

Table 1
Observed and expected frequencies of facial surfaces exhibiting both calculus and caries where each surface has been taken as the unit of analysis.

Facial surface tooth type	Calculus+ & caries+ Observed	Expected	p (Fisher's exact)	Facial surface tooth type	Calculus+ & caries+ Observed	Expected	p (Fisher's exact)
18	1	0.5	1.00	48	0	0.4	1.00
17	0	1.1	0.52	47	0	0.5	0.99
16	2	1.6	1.00	46	0	0.5	1.00
15	0	0.3	1.00	45	0	0.1	1.00
14	0	0.6	0.92	44	0	0.8	0.76
13	0	0.4	1.00	43	1	1.6	0.93
12	0	0.1	1.00	42	0	0.3	1.00
11	0	0.1	1.00	41	0	0.3	1.00
21	0	0.1	1.00	31	0	0.4	1.00
22	0	0.1	1.00	32	0	0.4	1.00
23	2	0.5	0.11	33	1	2.0	0.70
24	0	0.6	0.91	34	0	1.7	0.34
25	1	0.5	0.95	35	0	0.5	0.95
26	1	1.4	1.00	36	0	1.4	0.43
27	1	1.5	1.00	37	0	1.2	0.49
28	1	0.7	1.00	38	0	0.5	1.00

Hypothesis 2
That supragingival calculus is as likely to occur on any surface of a given tooth whether or not any surface of that tooth has a caries lesion. For this analysis, we have included the occurrence of caries on the occlusal surfaces. Analyses were performed using Fisher's exact test, using each tooth as the unit of analysis.
    There was no association apparent between caries on any given tooth type and the occurrence of calculus on the same tooth, except in the 55- to 65-year-olds. In the latter cohort, a lower than expected number of individuals had caries concurrently present with calculus on the same tooth type: T17 (p=0.0014); T16 (p=0.0503); T26 (p=0.0029); and T27 (p=0.0016).

Hypothesis 3
That there is no association between caries experience of an individual and the number of surfaces exhibiting supragingival calculus in that individual (irrespective of the site at which these two characteristics occur). Pearson's correlation coefficients were calculated to estimate the extent of association between the number of surfaces exhibiting calculus and the number of surfaces with caries. In this analysis, occlusal surface caries was included in the calculation of the individuals's DS score.
    The mean number of surfaces with caries and those with calculus present are shown in Table 2. The table also shows the correlation coefficients for the association between the number of surfaces exhibiting supragingival calculus and the number of surfaces with caries. In all age groups a negative as-

Table 2
Mean number of sites per individual with supragingival calculus and with caries (standard diviations in brackets), and the
association between caries and calculus

Age	Supra gingival calculus	Caries (incl. occlusal surfaces)	r Calculus vs caries	N
15-24	3.42 (5.75)	0.49 (1.04)	-0.045	256
25-34	6.41 (8.85)	1.47 (2.44)	-0.092	216
35-44	8.43 (10.93)	2.15 (4.21)	-0.111	220
45-54	9.98 (10.16)	1.82 (2.62)	-0.116	216
55-65	11.98 (12.32)	3.42 (5.92)	-0.108	223
All	7.91 (10.20)	1.83 (3.73)	-0.021	1131

r (with age) 0.29   0.242
              p < 0.0001   p < 0.0001

sociation was apparent, although for each age group the association was statistically non-significant.
Hypothesis 4
That there is no association between the occurrence of supragingival calculus and caries occurring on the same surfaces
within individuals.
    Fig. 1 shows the frequency distribution of individuals in
whom there were sites on which both calculus and caries was
observed. 4.8% of individuals had such sites. The mean number
of sites (in such individuals) that were observed is compared
with the expected number (based on the marginal distributions)
in Fig. 2 according to age group. In the oldest age group the
observed and expected number were approximately equivalent
and, with declining age, the observed exceedeed the expected
by a considerable amount. The $\log_e$ odds ratios for each age
group are shown in Fig. 3, indicating a strong positive association between caries and calculus occurring on the same surfaces, being strongest in the youngest cohorts, but weak in
the oldest.

DISCUSSION

Given that these data originate from a population for whom access to dental care is minimal (17), it has been possible to
investigate the association between supragingival calculus and
caries witout undue fear that the relationship would be masked
by the effects of dental treatment. In a previous study we
have shown (albeit based on an analysis of cross-sectional

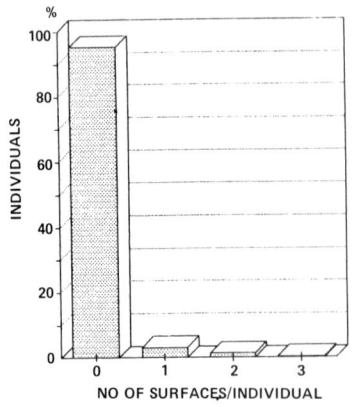

Fig. 1. Frequency distribution of surfaces per individual having both calculus and caries present.

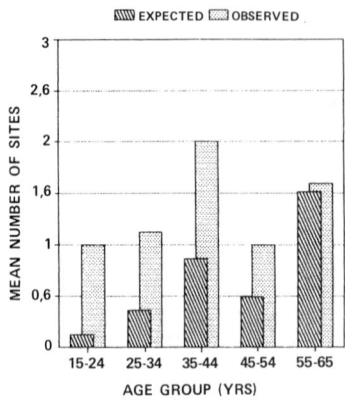

Fig. 2. Mean number of observed and expected sites with both calculus and caries present by age group.

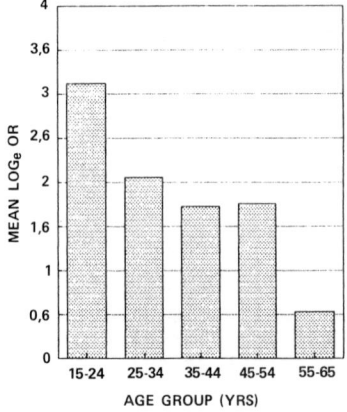

Fig. 3. Mean $\log_e$ odds for the association between calculus and caries occurring on the same surface, by age group.

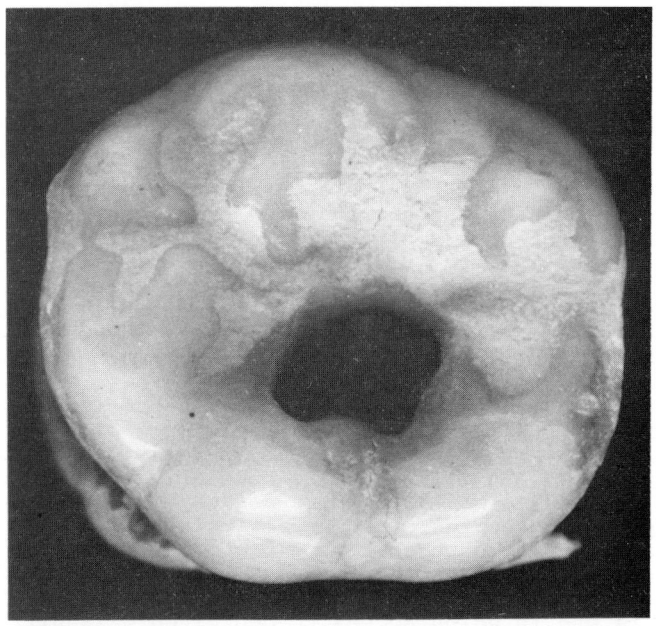

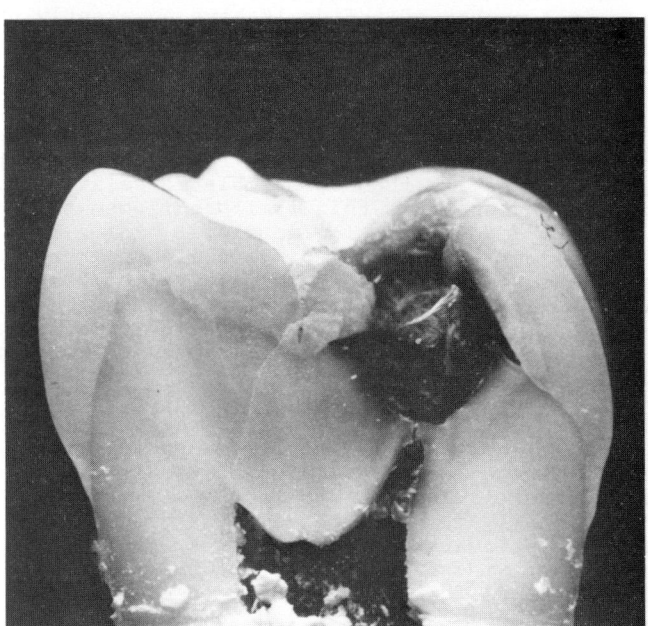

Figs. 4a & b. Molar with deep occlusal caries lesion and calculus throughout the occlusal fissure system.

data) that caries progresses slowly with age (18) in the context of poor oral hygiene in a population in whom calculus deposition apparently increases with increasing age (19). Although both caries and calculus were thus, positively associated with age (Table 2), both parameters were found to be negatively associated with each other when the relationship between individuals was considered.

It is not surprising that the strength of associations observed here were weak. In part this may be accounted for by
the cross-sectional nature of this study. However, given that
the proclivity of a site (or individual) to develop caries is
the result of the influence of multiple determinants (salivary
flow rate, buffer capacity, pH, nature of microbial deposits,
etc.) it would be unreasonable to expect that any one single
determinant (e.g. calculus) would account for much of the variance in caries experience.

The finding of a tendency for a negative association between calculus deposition and caries experience of the individual where analyses were performed across individuals was to
have been expected. At first glance, however, it may seem
surprising that there was a positive association between calculus formation and caries experience within individuals. Although a number of hypotheses may be proposed to explain such
findings, our view is that the findings were a product of the
very nature of the population studied - that is to say, the
population is one in which access to dental care is difficult.
Although we know that caries progresses slowly in this population (18), a small number of indiduals have apparently developed advanced lesions that may be sufficiently painful to
result in the individuals avoiding either cleaning of, or
chewing on or around, that site. Under such circumstances the
entire region surrounding the affected site will be stagnant
and will accumulate calculus over time. As unstimulated saliva
is known to be more supersaturated with respect to both hydroxyapatite and fluorapatite than is stimulated saliva (2), it
is tempting to speculate that this will predispose the non-
functional side to deposition of minerals.

This apparent contradictory finding (the concurrent occurrence of deep caries lesions and supragingival calculus - see
examples in Figs. 4a & b) to us principally reflects the apparently wide variations in physiological and physico-chemical
conditions that can prevail at the same time within relatively
small distances in the mouth. The oral cavity should not,
therefore, be considered as a single homogeneous unit in so
far as the factors which determine calculus deposition are
concerned. This is, for example, equally apparent from the
nature of the distribution of calculus in the mouth reported
in previous studies (13,20,21). The fact that supragingival
calculus predominantly forms in those areas close to the openings of the major salivary glands indicates, moreover, that
distinct gradients exist in the degree of supersaturation of
saliva with respect to the various minerals over relatively
small distances. If this is considered in the context of the
illustrative studies of Jenkins and Krebsbach (22) in which
they followed, over time, the pattern of distribution of char-
coal particles within the mouth, then there appears to be con-

siderable evidence to support the concept of the existence of a number of gross and almost discrete compartments within the mouth. Such a concept finds support also in the findings related to glucose and fluoride clearance from the mouth (23,24).

It should be appreciated that dental plaque is an extremely heterogeneous deposit. Its composition, both in terms of the microflora, intermicrobial matrix, and distribution of ion concentrations, can vary over rather small distances. The significance of this is probably best demonstrated by the fact that it is possible histologically to find, along a single root surface, focal areas of mineral dissolution separated by only a few microns from areas of unaffected surface cementum, although the entire surface is completely covered by plaque (25). The findings here emphasize, therefore, that the nature of the relationship between calculus deposition and caries formation has to be studied at the micro-level.

REFERENCES

1. McCann, H.G. (1968) in Art and science of dental caries research, (ed. Harris, R.S.) pp. 55-73. Academic Press, New York.
2. Larsen, M.J. (1975) Enamel solubility, caries and erosion, Thesis, Royal Dental College, Aarhus.
3. Driessens, F.C.M., Borggreven, I.M.P.M., Verbeeck, R.M.H., van Dijk, J.W.E. and Feagin, F.F. (1985) J. Periodont. Res. 20, 329-336.
4. Pierce (1982) Caries Res. 16, 460-471.
5. Pierce (1984) Caries Res. 18, 103-110.
6. Stones, H.H., Lawton, F.E., Bransby, E.R. and Hartley, H.O. (1950) Brit. Dent. J. 29, 199-203.
7. James, P.M.C. (1963) Proc. R. Soc. Med. 56, 620-624.
8. Marthaler, T.M. and Schroeder, H.E. (1966) Helv. Odont. Acta 10, 120-130.
9. Sutfin, L.V., Sweeney, E.A. and Ascoli, W. (1970) J. Dent. Res. 49, 772-775.
10. Jacobsen, M. and Kesel, R.G. (1950) J. Dent. Res. 29, 364-374.
11. Little, M.F., Wiley, H.S. and Dirksen, T.R. (1969) J. Dent. Res. 39, 1151-1162.
12. Davies, G.N. (1956) 35, 734-741.
13. Ainamo, J. (1970) 66, Finska Tandläkarsällskapets Förhandlingar 305-366.
14. Björby, Å. and Löe, H. (1967) J. Periodont. Res. 2, 76-77.
15. Berkey, C.S., Douglass, C.W., Valachovic, R.W. and Chauncey, H.H. (1988) Community Dent. Oral Epidemiol. 16, 83-90.
16. Kapur, K.K., Glass, R.G., Loftus, E.R., Alman, J.E. and Feller, R.P. (1972) Aging Hum. Dev. 3, 125-137.
17. Manji, F., Baelum, V. and Fejerskov, O. (1988) J. Dent. Res. 67, 496-500.
18. Manji, F., Fejerskov, O. and Baelum, V. (in press - 1988) Caries Res.
19. Baelum, V., Fejerskov, O. and Manji, F. (1988) J. Clin. Periodontol. 15, 445-452.

20. Baelum, V. (1987) Scand. J. Dent. Res. 95, 221-228.
21. Alexander, A.G. (1971) J. Periodont. 42, 21-28.
22. Jenkins, G.N. and Krebsbach, P.H. (1985) Arch. Oral Biol. 30, 697-699.
23. Weatherell, J.A., Strong, M., Robinson, C. and Ralph, J.P. (1986) Caries Res. 20, 111-119.
24. Britse, A. and Lagerlöf, F. (1987) Arch. Oral Biol. 32, 755-756.
25. Nyvad, B. and Fejerskov, O. (1987) in Dentine and dentine reaction in the oral cavity, (eds. Leach, S.A. and Thylstrup, A.) pp 165-179. IRL Press, Oxford, UK.

G.K.Stookey
R.D.Jackson
B.B.Beiswanger
K.R.Stookey

Clinical efficacy of chemicals for calculus prevention

Indiana University School of Dentistry, Oral Health
Research Institute, 415 Lansing Street, Indianapolis,
IN 46202, USA

Abstract

A wide variety of chemical compounds for the prevention of
calculus formation have been investigated during the past 30 years.
These compounds have included antimicrobial agents, enzymes, and a
number of crystal growth inhibitors. Clinical assessment methods
have involved gravimetric measurements of intra-oral calcified
deposits on plastic films (i.e., Mylar strips) and controlled
clinical trials utilizing various population groups and usage
conditions. The objective of this presentation is to review the
results of these various approaches for preventing calculus
formation. Particular emphasis will be given to the results of
controlled clinical trials of agents which are presently available
for use by the general population; these agents include
diphosphonates, zinc salts, and soluble pyrophosphate systems.
Available clinical evidence clearly documents the ability of these
compositions to reduce calculus formation in the population.

Recent Advances in the Study of
Dental Calculus

Introduction
     Dental calculus deposits are present in a significant proportion
of the adult population. Marshall-Day et al. (1), in 1955, observed
calculus in 66% of the adult population. In 1982 the results of a
survey in Iowa (USA) found calculus in 45% of the adult population
(2). More recently, a national survey conducted by the U.S. Public
Health Service found that only 16% of employed adults and 11% of
senior citizens were free from calculus deposits (3).
     While dental calculus per se is no longer considered to be a
primary etiologic factor in the development of periodontal disease
(4), calculus is accepted as a secondary contributing factor, and
the removal of these deposits for both cosmetic and therapeutic
reasons comprises a large percentage of the workload in most dental
practices. In a national survey of U.S. adults, 16.5% of employed
adults and 19.7% of senior adults cited a dental prophylaxis as the
main reason for their last dental visit (3).
     For many years a number of researchers have devoted their
professional skills to the discovery of various therapies designed
to prevent or inhibit the development of these deposits. While the
results of many of the earlier clinical trials were based on
subjective clinical observations rather than the currently accepted
investigative methods and procedures, they provided some insight
into how the development of calculus deposits could be controlled.
The purpose of this paper is to review some of the chemical agents
which have been reported to inhibit or suppress the development of
dental calculus in human clinical trials.
     A discussion of chemotherapeutic agents designed to inhibit
dental calculus formation needs to first recognize that there are
two different types of dental calculus, as defined by the area of
occurrence: supragingival and subgingival. This review is
restricted to investigations of supragingival calculus.

Enzymes
     Early attempts to develop a means of removing dental calculus
from the teeth centered on the use of enzymes. This approach was
based on the premise that calculus was believed to contain
approximately 30% water and organic matter and that the use of an
enzyme would cause dissolution of the organic matter in calcified
deposits. Table 1 summarizes the results of clinical trials with
enzyme preparations.
     One of the earliest reports was by Stewart (5). Stewart believed
that salivary mucins were essential in the attachment of calculus to
the tooth surface and destruction of these proteins would lead to a
reduction in calculus formation. His clinical impressions and
observations in three subjects led him to believe that mucinase was
beneficial in retarding the growth of calculus when used in an
ammoniated dentifrice.
     Aleece and Forscher (6) performed a small clinical trial to
determine the effectiveness of mucinase dentifrices. The
investigators explained that the term mucinase was not related to

Table 1. Studies Involving Enzyme Preparations

Study	Agent	No.	Length	% Reduction
Stewart (1952)[5]	Mucinase	3	Varied	Decrease
Aleece & Forscher (1954)[6]	Mucinase	11	6 months	Decrease
Jensen (1959)[7]	Pancreatic Enzymes	134	6 weeks	60
Ennever & Sturzenberger (1961)[8]	Pancreatic Enzymes	19	8 weeks	24
Packman et al. (1963)[9]	Pancreatic Enzymes	47	6 months	15
Packman et al. (1963)[9]	Fungal	47	6 months	41
Harrison et al. (1963)[10]	Proteolase	211	6 months	22
	Amylolase			7
	Cellulase			3 (N.S.)

one specific enzyme, but rather to a group of compounds which have the ability to hydrolyze mucin-like material. Eleven subjects were used to test three dentifrices containing 3% "Prolase 100", "Mylase L-1", or papain. Powders were used as vehicles for the test agents to maintain stability of the enzymes. After six months of home use, the panelists using "Mylase L-1" had less clinically observable dental calculus. The investigators surmised that the amylolytic activity of Mylase was greater than the proteolytic activity of Prolase.

Jensen (7) investigated the use of dehydrated pancreas (Viokase) and dispersant lactose. Viokase contains all the enzymes available in the pancreas and, as such, Jensen believed that the benefits could not be attributed to one particular enzyme. The clinical trial involved 134 panelists who applied their assigned treatments once daily for 10-15 minutes. The results indicated that Viokase reduced dental calculus accumulation by 60% after six weeks of use. Studies of the time needed to complete a dental prophylaxis indicated that the group using Viokase required 24.8 minutes versus 36.2 minutes for those panelists using the placebo powder, or a decrease of 32% per dental prophylaxis.

Viokase has also been incorporated into a chewing gum base by Ennever and Sturzenberger (8). In preliminary research they found that Viokase reduced calculus accumulation, but most panelists objected to the taste, texture, and odor of the compound. A chewing gum base was chosen to allow adequate contact time between the enzymes and the calculus deposits, as well as to provide a vehicle which would be compatible with the enzymes. A panel of 19 subjects was given a dental prophylaxis. The panelists were scored using the Calculus Surface Index (C.S.I.) following an eight-week pretest period and were given their assigned product (active or control). Each panelist was instructed to chew one stick of their assigned product five times a day for five minutes each time. At the end of the eight-week test period, the group using the chewing gum containing Viokase had 24% less calculus when compared to the control group.

A study by Packman et al. (9) also tested the addition of enzymes to a chewing gum base as a means of decreasing calculus accumulation. Enzymes derived from fungi were incorporated into a chewing gum base and were compared to pancreatic enzymes in the same type of chewing gum base. Forty-seven panelists were randomly assigned to one of the two products. After six months the results indicated that the fungal enzymes had reduced calculus accumulation by 40.5% as compared to 14.5% by the pancreatic enzymes.

Harrison et al. (10) incorporated several fungal enzymes into dentifrices. One was considered to be highly proteolytic, one was amylolytic, and the other was a cellulase. The subjects using the proteolytic enzyme preparation exhibited a 22% reduction in calculus while the amylolytic preparation reduced calculus by 7%. The cellulase preparation was not found to be clinically effective.

Chelating Agents

Chelating agents are known to sequester and dissolve calcium salts by forming stable, soluble calcium complexes. Early investigations of sodium hexametaphosphate found that this compound could reduce or destroy kidney stones, and it was known to be useful in preventing the accumulation of calcium carbonate scale on industrial equipment. These factors led Kerr and Field (11) to investigate the use of sodium hexametaphosphate to prevent calculus accumulation (Table 2). A panel of ten subjects, known to be heavy calculus formers, was instructed to use a saturated solution of sodium hexametaphosphate as a mouthrinse and as a dentifrice once daily for one minute each. The panelists were then followed for varying lengths of time ranging from a few weeks to several months. At the end of the trial, the authors concluded that the use of sodium hexametaphosphate as a mouthrinse and as a dentifrice was effective in inhibiting calculus formation for six weeks or longer. Laboratory investigations confirmed this result, but the authors also noted that the compound would cause complete dissolution of silicate restorations. An in vitro study by Warren et al. (12) found that a saturated solution of sodium hexametaphosphate caused a

Table 2. Studies Evaluating Chelating Agents

Study	Agent	No.	Length	% Reduction
Kerr & Field (1944)[11]	$(Na_4(PO_{36})$	10	Varied	100
Grossman (1954)[13]	Ex347	31	Varied	Big Red. in 19 Subjects
Gunson (1955)[14]	Ex347	368	12 weeks	100% in most Subjects

severe reduction of the hardness of cementum in teeth exposed for 24 hours to the solution.

In 1954 Grossman published a paper describing the use of another chelating agent referred to as Ex347 (13); these data are also presented in Table 2. Based on his earlier in vitro study, Ex347 had been found to be effective in removing calculus, and it did not appear to have any deleterious effects on enamel. The agent was tested in a group of 31 panelists who were known to be calculus formers. The solution was applied twice daily for one minute and was allowed to remain without rinsing. No control solution was used and the results were based upon clinical observations alone over test periods lasting up to six months. Of the 31 panelists, 19 showed a marked reduction in the amount of calculus formed while eight panelists had slight reductions.

Gunson (14) also investigated the efficacy of Ex347 in a large clinical trial of 368 male prisoners. Each panelist was supplied with the Ex347 solution and a multi-tufted toothbrush. No control group was used in the investigation. The panelists were instructed in the proper toothbrushing procedure and were told to use their product twice daily. After three months of use, Gunson observed that 95% of the panelists had total removal of existing calculus and had total inhibition of new calculus and stain formation. The remaining five percent had evidence of partial dissolution of the existing calculus and stain, and no new deposits were noted. These results were based totally on clinical impressions; no recognized dental calculus indices were employed.

Although Gunson reported that only a few of the panelists had any untoward effects from the use of Ex347, Leung reported that the solution was, in fact, detrimental. When extracted teeth were submerged in solutions comparable to those recommended by the manufacturers, enamel damage was evident (15).

Table 3. Clinical Studies of Antimicrobial Agents

Study	Agent	No.	Length	% Reduction
Dossenbach & Muhlemann (1961)[16]	Pencillin	20	1 Week	0
Muller et al. (1962)[17]	Cetylpyridinium Chloride	10	7 Days	0
Muller et al. (1962)[17]	Salt Mixture*	10	7 Days	N.S.
Kristoffersen (1963)[18]	Salt Mixture*	134	30 Days	0
Sastri et al. (1966)[19]	Salt Mixture*	20	Varied	0
Westerholm (1971)[20]	Salt Mixture*	20	3 Years	Sig.
Volpe et al. (1969)[21]	CC 10232	36	12 Weeks	69-80
Volpe et al. (1970)[22]	CC 10232	72	9 Months	33
Stallard et al. (1969)[23]	CC 10232	60	7 Days	70-91
Rokita et al. (1975)[24]	CC 10232	5	2 Weeks	Some
Cancro et al. (1972)[25]	Chlorhexidine	18	2 Days	9 & 40
Löe et al. (1971)[26]	Chlorhexidine	6	15 Days	100
Löe et al. (1976)[27]	Chlorhexidine	120	2 Years	Sig. Incr.
Hirsch & Leclercq (1978)[28]	Chlorhexidine	21	14 Days	16
Hirsh & Leclercq (1978)[28]	F-Hexidine **	20 / 18	14 Days / 21 Days	11 / 8 (N.S.)
Albertini et al. (1983)[29]	Sodium Benzoate	60	30 Days	+ 7 (N.S.)

* Mixture of ascorbic acid, sodium percarbonate, and copper sulfate
** Difluorhexidine - difluorhydrate (DFDFH)

Antimicrobial Agents

Although it has been shown that microorganisms are not essential for the development of dental calculus, it has been suggested that they may contribute to calculus formation. Therefore, several investigators have tried to determine if suppressing the numbers of microorganisms present in the oral cavity could influence the rate of deposition and formation of dental calculus. Table 3 summarizes the results of clinical trials evaluating the use of antimicrobial agents for calculus prevention.

Dossenbach and Mühlemann (16) studied the effects of penicillin in humans as a means of controlling calculus deposition using the Mylar strip model in 20 dental students. Penicillin was chosen as the antibiotic because of its known efficacy to gram-positive cocci and rods which predominate in early dental plaque. Sodium penicillin lozenges (5000 I.U.) were administered ten times daily to provide a total of 50,000 units per day. This treatment was begun one day prior to the attachment of the Mylar strips and was concluded after five days. Results of the use of the penicillin lozenges showed no decrease in the formation of calculus deposits.

Muller et al. (17) tested two oral antiseptics on ten students using the Mylar strip method with seven-day test periods. Two systems were tested against a negative control: one contained cetylpyridinium chloride (CPC) and dequalinium acetate (Micrin); the other contained a salt mixture consisting of ascorbic acid (100mg), sodium percarbonate (70mg) and copper sulfate (0.2mg) (Ascoxal). The control and CPC were supplied as mouthrinses and the salt mixture was supplied in tablet form to be dissolved in 20 ml water before use. The students were told to rinse with 20 ml of their assigned solutions 3 times per day for two minutes each time and again after toothbrushing at least twice per day. At the end of the study, the investigators reported no clinically apparent differences between the three rinses. The gravimetric assessment showed no statistical differences between the control and CPC groups; the salt mixture had an inhibitory effect which was significant at $P<0.01$, but it had little effect on calculus in "rapid formers".

Similar results were reported by Kristoffersen (18) using a panel of 134 patients in a double-blind study comparing the same salt mixture and a placebo. No statistical differences could be found in calculus formation at the end of 30 days. Sastri and Yuktanandana (19) also found no difference in a panel of 20 periodontal patients when similarly comparing the use of this salt mixture to a placebo solution. However, a study by Westerholm (20) showed a significant reduction in calculus (C.S.I.) among patients assigned to the use of the product as part of their homecare routine during a three-year test period.

In a series of studies performed by Volpe et al. (21), a compound referred to as CC 10232 was found to have an effect on the deposition of dental calculus. The compound was described as a macrolide antibiotic obtained from the fermentation of a novel strain of Streptococcus caelestis. It showed strong activity against a variety of gram-positive bacteria, but not gram-negative bacteria or yeasts. The compound was tested in a twelve-week clinical trial involving 36 institutionalized adult subjects between the ages of 21 and 50 years. The panelists were stratified by calculus present at the baseline examination and then were randomly assigned to the test (0.01% CC 10232) or a placebo mouthrinse group. The supervised program involved toothbrushing with a placebo dentifrice three times per day followed by rinsing with 15 ml of

Table 4. Clinical Studies Using Various Agents for Calculus Prevention

Study	Agent	No.	Length	% Reduction
Belting & Gordon (1966)[30,31]	30% Urea	68	2 Weeks	36
Shapiro et al. (1973)[32]	11% Urea Perox.	80	6 Months	N.S.
Son & Mühlemann (1971)[33]	Urease Inhibitor	10	1 Week	N.S.
Dossenbach & Mühlemann (1961)[16]	Ricinoleate	14	1 Week	100
Swenson & Bixler (1967)[35]	Iontophoresis	92	1 Year	N.S.
Putt & Kleber (1988)[36]	Al Citrate	381	3 Months	16-22
Tsuprova (1986)[37]	Ammonium Chloride	45	9 Months	Decrease

their assigned mouthrinse for one minute. The results indicated that the group using CC 10232 had 64% less calculus at the end of six weeks than the control group. At 12 weeks two examiners found approximately 68.5% and 80% less calculus in the active group.

In 1970 Volpe reported the results of a larger, nine-month clinical trial involving 72 patients (22). Examinations were performed following three, six, and nine months of product usage. Thirty-six panelists were assigned to use a 0.01% CC 10232 mouthrinse and a nontherapeutic dentifrice, while the remainder used a control rinse and the same dentifrice. The active group had a 38% reduction in calculus at three months, 50% at six months, and a 33% reduction in calculus at the conclusion of the investigation. The investigators believed the decrease in the effectiveness of the active rinse at nine months was due to a decrease in calculus accumulation in the control group. Reductions in calculus accumulation have also been reported by Stallard et al. (23) and Rokita et al. (24) with the use of a mouthrinse containing 0.01% CC 10232. Stallard et al. reported a 70% to 91% decrease in calculus compared to a placebo. Rokita et al. observed a reduction in calculus formation in four of the five subjects using the rinse.

Cancro et al. (25) studied the use of a 0.1% chlorhexidine gluconate mouthrinse to inhibit early calculus formation using the Mylar strip model. Eighteen panelists were involved in two independent trials. The first was a 20-day crossover investigation;

the second was a 20-day study involving a linear design. The results of the linear study indicated that chlorhexidine gluconate caused a significant 40% reduction in the development of early calculus deposits. In the crossover design, the use of chlorhexidine was not found to be effective; however, the authors noted an order-of-use effect with reductions increasing with time. Löe et al. (26), after a study involving six dental students in two 15-day trials, concluded that the use of chlorhexidine gluconate applied as a mouthrinse (0.2%) or topically (2.0%) inhibits the formation of supragingival calculus. However, in 1976 Löe et al. reported that the use of a 0.2% solution of chlorhexidine increased the deposition of supragingival calculus over a two-year period (27). Using a gravimetric procedure, Hirsh and Leclercq (28) observed a 16% decrease in calculus formation associated with the use of a 0.1% chlorhexidine mouthwash during a 14-day period.

Hirsh and Leclercq (28) also used a difluorohexidine mouthwash and observed a significant 11% decrease in calculus after 14 days, although a numerical increase was observed after 21 days. Albertini et al. (29) evaluated a dentifrice containing sodium benzoate and observed no effect upon calculus formation as compared to a placebo.

Other Agents

Table 4 summarizes the results of trials with a variety of other agents evaluated for calculus prevention. Belting and Gordon (30, 31) tested the efficacy of a dentifrice containing 30% urea. A panel of 68 subjects was divided into two groups and instructed to use their assigned dentifrice twice daily (29). One group used the control dentifrice for 28 days, while the other used the control for 14 days followed by the urea dentifrice for the remaining 14 days.

At the end of the four-week trial, the group using the active product had a 36% decrease in calculus formation as determined by the amount of calcium present on a resin strip attached to one of the mandibular incisors. However, Shapiro et al. (32) found that the use of a gel containing 11% urea peroxide actually caused an increase of calculus deposition in 80 panelists who were instructed to rub a small amount of their assigned gel on the teeth for two minutes twice daily following toothbrushing.

Son and Mühlemann (33) tested the effects of acetohydroxamic acid, a potent urease inhibitor, on the development of dental calculus. Previous research by Mandel and Thompson (34) had shown high levels of urea in the submaxillary saliva of heavy calculus formers. It was believed that an inhibitor of urease would inhibit the formation of urea and thus reduce calculus formation. Using the Mylar strip model in ten dental students with two seven-day test periods, no significant differences were found between the use of the acetohydroxamic acid solution and the water control.

Dossenbach and Mühlemann selected sodium ricinoleate for use as a positive control on the basis of in vitro reports of the effectiveness of similar agents (16). The trial employed a Mylar strip model with 14 dental students as subjects. The ricinoleate was administered as a 4% topical solution applied three times per day. In addition the subjects used a 0.1% ricinoleate mouthrinse five times per day. This continuation treatment totally prevented calculus deposition onto the Mylar strips during the one week test period.

In 1967 Swenson and Bixler reported the results of an investigation to determine the effect of iontophoresis on the development of dental calculus (35). The one-year double-blind investigation involved 92 adults in a private dental practice. The panelists were randomly assigned to use an Ion toothbrush or a control toothbrush which could not generate electrical current. The panelists were instructed to use their assigned brush with a nontherapeutic dentifrice which was supplied. They were then examined during their regular recall appointments. At the end of the study, no significant differences were found between the two groups.

A study by Putt and Kleber (36) was designed to test the efficacy of chewing gum containing aluminum citrate (AlCt) to inhibit calculus formation. A total of 381 adult panelists took part in the three-month trial. The panelists were stratified into four groups based on their calculus scores following a three-month pretest period. The aluminum citrate chewing gum was used with either a placebo dentifrice or one containing soluble pyrophosphate. Subjects who received the gum were instructed to chew three or more pieces per day for at least five minutes each. At the end of the test period, the groups using the aluminum citrate gum had reductions in calculus scores of 16.2 to 21.6%.

A tooth powder containing 2% ammonium chloride was evaluated in a clinical practice situation by Tsuprova (37). After periods up to nine months, a reduction in calculus formation was observed.

Crystal Growth Inhibitors

Turesky et al. (38) reported the results of a study in which the ability of the chlormethyl analogue of Victamine C to inhibit calculus formation was evaluated using the Mylar strip model (Table 5). A silicone rubber tray was utilized to isolate the mandibular teeth and allow maximum contact of a one percent solution with the tooth surfaces in 11 male subjects. The five-minute treatments were performed three times daily for an eight-day period. This same regimen was repeated with distilled water as the control. The average reduction was 46.8%.

To rule out the possibility that the taste of the active agent may have influenced the rate of calculus formation by increasing salivation, the study was repeated with quinine sulfate as the control agent (39). Quinine sulfate was selected due to its similarity in taste to the test agent and because it does not inhibit crystal seeding. Five males were selected for the second test and the study was performed in essentially the same manner as the previous study. The average reduction in calculus weights was 70.9%.

Mühlemann et al. (40) studied the effects of sodium etidronate, a disodium salt of ethane-1-hydroxy-1,1-diphosphonate (EHDP), on the accumulation of dental calculus in 21 dental students ages 22 to 30. He used the seven-day Mylar strip model to compare the effects of a 0.125% solution of EDHP to a control and to two rinses, identified as CAF and CA, which contained a calcium complexing agent with and without fluoride. The results indicated that EHDP depressed the formation of calculus deposits by 39% without interfering with the degree of mineralization. This difference was 49% in rapid calculus formers but in slow calculus formers, EHDP had little effect (16%) on deposit weight although the deposits were less mineralized.

Sturzenberger et al. (41) investigated the efficacy of a dentifrice containing 3% sodium etidronate (EHDP) as a means to inhibit calculus formation by affecting crystal growth. Sixty-four subjects ranging in age from 13 to 64 years were prescreened and only those persons receiving a score of 6 or greater by the Calculus Surface Index (C.S.I.) were allowed to continue. The panelists were instructed to use their assigned dentifrice (active or control) in their normal manner, but at least once per day, following a dental prophylaxis. Examinations were then performed after eight weeks and six months using three different indices (C.S.I., C.S.S.I., V.M.I.). To determine the effect of sodium etidronate on existing calculus deposits, the group using the control dentifrice during the preceding six-month period was divided into two groups balanced on their calculus scores. These two groups were then randomly assigned to the active dentifrice (3% sodium etidronate) or continued with the control dentifrice for an additional eight weeks. The results indicated that the group using the active dentifrice had significant reductions in calculus accumulation of 36% to 51% at eight weeks and 46% to 66% at six months, depending on the index used. In the second phase of the study, those members of the control group who used the 3% sodium etidronate dentifrice were found to have significantly less calculus (25%) than those who continued with the control dentifrice.

In 1972 Conroy et al. reported the results of a second study of a 3% sodium etidronate dentifrice using a similar study design but without the post-test phase (42). In this trial 0.22% sodium fluoride was also added to the dentifrice. At the end of the six month trial, calculus formation was reduced by 28% to 51%, depending on the index utilized.

Suomi et al. (43) performed an 18-month clinical trial to compare the effects of a dentifrice containing 3% sodium etidronate and 0.22% sodium fluoride to a placebo dentifrice which did not contain either of these ingredients. The parameters which were measured included both supragingival and subgingival calculus. Examinations were performed at baseline, 6, 12, and 18 months. The 244 panelists were between 20 and 57 years of age and received a dental prophylaxis prior to receiving their test products. No toothbrushing instructions or recommendations were given for

Table 5. Studies of Crystal Growth Inhibitors: Victamine C and Diphosphonate

Study	Agent	Conc.	No.	Length	% Reduction
Turesky et al. (1967)[38]	Victamine C	1.0%	11	8 Days	47
Gilmore et al. (1968)[39]	Victamine C	1.0%	5	21 Days	71
Mühlemann et al. (1970)[40]	Diphosphonate*	0.125%	21	1 Week	39
Sturzenberger et al. (1971)[41]	Diphosphonate*	3.0%	64	6 Months	46–66
Conroy et al. (1972)[42]	Diphosphonate*	3.0%	391	6 Months	28–51
Suomi et al. (1974)[43]	Diphosphonate*	3.0%	244	18 Months	42
Videmari (1977)[44]	Diphosphonate*	?	106	6 Months	60
Herforth et al. (1977)[45]	Diphosphonate*	1.0%	50	3 Months	34–41
König (1978)[46]	Diphosphonate*	0.5%	43	2 Months	46
Laturnus (1974)[47]	Diphosphonate*	0.1%	39	3 Months	44
Seichter et al. (1979)[48]	Diphosphonate*	0.5%	192	2 Months	Sign.
Kremers et al. (1980)[49]	Diphosphonate*	1.0%	26	4 Months	N.S.
Halbritter (1973)[50]	Diphosphonate*	0.5%	24	2 Months	48
Herforth (1976)[51]	Diphosphonate*	1.0% 0.1%	14 14	7 Days 7 Days	56 57
Seichter & Herforth (1980)[52]	Diphosphonate-DPD	0.5%	54	22 Weeks	22
Seichter & Laturnus (1980)[53]	Diphosphonate-PPT	1.0%	90	9 Months	Sign.
Cassese et al. (1980)[54]	Diphosphonate-PPT	1.26%	60	3 Months	38
Bubani (1980)[55]	Diphosphonate-AHP	1.15%	40	8 Weeks	23
Nilsson & Holm (1986)[56]	Diphosphonate-AHP	1.0%	?	4 Months	30

* Sodium etidronate, EHDP

frequency of use. The results showed that at each examination the test group had significantly less supragingival calculus (P<0.01) than the control. At 18 months the group using the active dentifrice had a 42.1% reduction in calculus scores as compared to the controls. Reductions in subgingival calculus were also observed, but only the results at the 12-month examination were statistically significant (P<0.05). However, silicate restorations in subjects in the EHDP group were found to have a lower degree of marginal adaptation than those in the control group, indicating that EHDP may have had a deleterious effect upon silicate.

In a longitudinal, placebo-controlled study, Videmari (44) similarly evaluated a dentifrice containing diphosphonate provided as sodium etidronate (EHDP). Following a six-month usage period, the experimental dentifrice resulted in a 60% decrease in calculus formation.

Herforth et al. (45) also investigated the use of EHDP as a means of controlling dental calculus. In a double-blind clinical trial involving 50 patients, they found that the use of 1% EHDP in a dentifrice reduced calculus by 34.1% when scored with the V.M.I. and by 41.1% using the C.S.I. The use of EHDP was also found to cause a 20.3% reduction in mineralization of the dental calculus which was formed. König (46) evaluated a silica dentifrice containing 0.5% sodium etidronate and observed a significant reduction in calculus formation of 45.6% after a two-month test period. Laturnus (47) similarly evaluated an essentially identical dentifrice containing 0.1% sodium etidronate and reported a calculus reduction of 43.7% after three months. Seichter et al. (48) also reported a significant reduction in calculus formation with a dentifrice containing 0.5% sodium etidronate. No differences in calculus formation were found in the study reported by Kremers et al. (49), which involved 26 panelists.

The use of sodium etidronate in a mouthwash for the prevention of calculus has also been investigated. Halbritter (50) reported that twice daily rinsing with a solution containing 0.5% sodium etidronate resulted in a 48% decrease in calculus formation after 2 months. Using the seven-day Mylar strip model, Herforth (51) observed reductions in calculus formation of 56.2% and 56.8% following the use of rinses containing 1.0% and 0.1% sodium etidronate, respectively.

A number of other types of diphosphonates have also been evaluated. Seichter and Herforth (52) evaluated a dentifrice containing 0.5% diphosphono-propane-dicarboxylic acid (DPD). Unsupervised use of this dentifrice for periods of 11 and 22 weeks resulted in reductions of calculus formation of 15.8% and 21.5%, respectively. Seichter and Laturnus (53) reported a significant reduction in calculus formation associated with the use of a dentifrice containing 1% 1-phosphono-propane-1,2,3-tricarboxylic acid (PPT) during a nine-month test period. Cassese and coworkers (54) evaluated a dentifrice containing 1.26% PPT in a three-month trial. As compared to baseline scores, the placebo dentifrice resulted in a 9.5% decrease in calculus formation while the experimental dentifrice gave a 38.2% decrease in calculus. Bubani (55) evaluated a dentifrice containing 1.15% azacycloheptane-2,2-disphosphonic acid (AHP); after an eight-week test period a 23% decrease in calculus formation was observed. Similarly, Nilsson and Holm (56) observed a 30% reduction in calculus formation associated with the use of a dentifrice containing 1% AHP.

Mühlemann et al. (40) concluded that EDHP showed promise as a therapeutic means of reducing calculus accumulation, but they stressed that further research was desirable due to its reported effect on fluoride uptake by hydroxyapatite and that interference with the remineralizaton potential of dental plaque may increase its pathogenicity. However, recent reports by Arends and Dijkman (57) and Koch et al.(58) indicate that the addition of diphosphonate to dentifrices containing sodium fluoride does not inhibit the remineralization process nor clinical caries formation.

Crystal Growth Inhibitors: Zinc Systems

Although zinc salts have been found to have low antimicrobial activity, they have been shown to inhibit plaque and calculus formation in humans. Early investigations by Hanke (59) suggested that heavy metals may be useful in controlling dental plaque without affecting the composition of the oral flora. The results of studies using various zinc compounds are summarized in Table 6.

Based on preliminary data which suggested that zinc ions could disrupt the nucleation of calcium phosphate and thus inhibit precipitation, Picozzi et al. (60) investigated the calculus inhibition effects of a mouthrinse containing 1.0% zinc phenolsulfonate and 0.125% zinc tribromsalan. Zinc tribromsalan is a broad spectrum gram-positive antibacterial agent commonly used in hygiene products such as soap. The clinical study involved 86 adult males who used a test or placebo rinse three times daily in conjunction with toothbrushing. Calculus scores (Volpe-Manhold Calculus Index) were reduced by 60.3% during the first three-month cycle and by 53.1% during the second test cycle. Pooled calculus weights were reduced by 77% and 85%, respectively, in the two test cycles in those patients using the test rinse. The investigators attributed the effectiveness of the mouthrinse to the bacteriostatic properties of the zinc tribromsalan and to the ability of zinc to interfere with the conversion of amorphous calcium phosphate to hydroxyapatite.

In comparing a sodium fluoride/zinc chloride dentifrice to a commercially available fluoridated dentifrice, Kohut and Grossman (61) found a 46.1% difference at the end of a three-month clinical trial. Sixty-seven panelists completed the investigation and calculus accumulation was assessed on the lingual surfaces of the mandibular anterior teeth using the Volpe-Manhold Index.

Schmid et al. (62) investigated a mouthrinse containing 0.2% zinc

Table 6. Studies of Crystal Growth Inhibitors: Zinc Systems

Study	Agent	Conc.	No.	Length	% Reduction
Picozzi et al. (1972)[60]	Phenolsulfonate Tribromsalan	1.0% 0.125%	86	3 Mon.	53-60
Kohut & Grossman (1986)[61]	Zn & NaF	?	67	3 Mon.	46
Schmid et al. (1974)[62]	$ZnCl_2$	0.2%	21	1 Wk.	38
Lobene et al. (1987)[63]	$ZnCl_2$	2.0%	57	6 Mon.	40
Rustogi et al. (1987)[64]	$ZnCl_2$	2.0%	107	3 Mon.	46-50
Kazmierczak et al. (1988)[65]	ZnCit.	2%	193	6 Mon.	32
Stephen et al. (1987)[66]	ZnCit.	0.5%	2316	3 Yrs.	30*
Lobene (1987)[91]	ZnCit.	0.5%	200	3 Mon.	11 (N.S.)

* % difference in prevalence

chloride utilizing the Mylar strip model in 21 dental students known to be calculus formers. The students were instructed to use 10 milliliters of their assigned test or placebo mouthrinse twice daily. The results indicated that the zinc chloride mouthrinse reduced the quantity of calculus formed in a seven-day period by 38%; however, the degree of mineralization was not altered.

Lobene et al. (63) compared the anticalculus effects of a dentifrice containing 2.0% zinc chloride, 0.22% sodium fluoride, and a silica abrasive (Prevent) to a commercially available non-fluoridated dentifrice with a silica abrasive (Pepsodent). The acceptable panelists were required to have six unrestored mandibular anterior teeth and a minimum calculus score of 3.0, as measured by the Volpe-Manhold Index. Following a three-month pretest period, 57 acceptable panelists were randomly assigned to one of the test groups. The patients were examined (VMI) at the end of three and six months. The three-month data indicated that the group using the zinc chloride dentifrice had a calculus reduction of 49.9%; after six months, the reduction was 39.5%.

A 2.0% zinc chloride dentifrice has also been reported to be more effective in reducing the accumulation of dental calculus when compared to a dentifrice containing 3.3% soluble pyrophosphate. Rustogi et al. (64) compared these two dentifrices in a double-blind, randomized clinical trial of three months duration. Panelists were selected following a three-month pretest period during which they used a placebo dentifrice, and those having a V.M.I. score of 3.0 or greater were allowed to continue. The 107 panelists were given their assigned dentifrice following a dental prophylaxis, and they brushed twice daily for one minute each time under supervision. At the end of the three-month test period, all the panelists were re-evaluated using the V.M.I. The results indicated that while both dentifrices reduced calculus accumulation compared to the pretest period, the zinc chloride dentifrice group developed 27.3% less calculus ($p<0.01$) than the group using the dentifrice containing pyrophosphate.

However, in a six-month study to compare dentifrices containing either 2% zinc citrate or 3.3% soluble pyrophosphate, Kazmierczak et al. (65) found the zinc citrate dentifrice to be comparable to the pyrophosphate dentifrice in preventing calculus. This study involved 193 participants who were randomly assigned to one of the three test products. Test scores (V.M.I.) indicated a 32.1% decrease in calculus accumulation for the zinc citrate group and a 21.4% decrease for the group using the pyrophosphate dentifrice. Further, Lobene (91) failed to observe any significant effect on calculus formation during a 3-month test period with a dentifrice containing 0.5% zinc citrate while a formulation containing soluble pyrophosphate exerted a significant beneficial effect of 51%.

In 1987 Stephen et al. published the results of a three-year investigation of the anticalculus effects of dentifrices containing 0.5% zinc citrate when combined with varying levels of sodium monofluorophosphate (1,000; 1,500; and 2,500 ppm F) versus dentifrices containing the same amounts of fluoride alone (66). Approximately 2,300 school students were available at the three-year examination. Calculus was assessed annually, but only the baseline and three-year data were reported. A modified version of the calculus prevalence index developed by Volpe was used. This index recorded the presence or absence of calculus on the mandibular incisors and the maxillary molars. At the conclusion of the study, the panelists using the dentifrices containing zinc citrate had 30% less calculus present than those using the control dentifrice.

Crystal Growth Inhibitors: Pyrophosphates

Early investigations by Fleisch and coworkers (67-69) showed that inorganic pyrophoshate can inhibit the precipitation of calcium phosphate and may also inhibit nucleation. Inorganic pyrophoshate is a naturally occurring substance found in plasma and saliva. Preliminary research by Vogel and Amdur (70) indicated that lower concentrations of pyrophosphate are found in the parotid saliva of calculus formers than in persons who do not form a large amount of calculus. Edgar and Jenkins (71) found this same direct relationship when studying the differences in composition of dental plaque between low and high calculus formers. Several investigations have shown that pyrophosphates have the ability to reduce calculus accumulation in the laboratory (72, 73) and in animals (73). As with the diphosphonates, there was concern at one time that the addition of pyrophosphates to an oral care product may inhibit the remineralization of carious lesions. However, studies by Lu et al. (74), Strang et al. (75), White and Faller (76), and Mellberg et al. (77) have shown that the incorporation of soluble pyrophosphates does not inhibit the anticaries benefits derived from fluoridated dentifrices in artificial, as well as naturally-occurring, carious lesions.

Table 7 presents the results of clinical trials evaluating soluble pyrophosphates for calculus prevention. A study in 1966 by Kinoshita and Mühlemann (78) evaluated 2.9% pyrophosphate in a troche using the seven-day Mylar strip model. A numerical decrease in calculus formation of 14.6% was observed, but this difference was not significant.

In 1985 the results of two clinical investigations (79, 80) were reported. In both investigations the test product was a dentifrice containing 3.3% soluble pyrophosphate as a mixture of tetrasodium pyrophosphate and disodium dihydrogen pyrophosphate in combination with 0.243% sodium fluoride in a silica base. In both studies the control dentifrice contained sodium fluoride in a silica base with no pyrophoshate.

The study by Zacherl et al. (79) included 418 calculus-formers 18-74 years of age. Calculus assessments used a modification of the V.M.I. which included the facial and lingual surfaces of all the teeth rather than just the lingual surfaces of the mandibular anterior teeth. At the conclusion of the two-month pretest period, the acceptable panelists were separated by age and sex, stratified by calculus scores, and assigned to one of two test groups. Following a dental prophylaxis, the subjects were instructed to use their dentifrice in their normal manner but at least once per day. The results after six months indicated that the group using the dentifrice containing pyrophosphate had 32% less calculus (p < 0.05) than the control group. The investigation by Mallatt et al. (80) involved 217 adults with the same design features as that described by Zacherl et al. After a two-month test period, the reduction in calculus was 26% for both examiners (p < 0.01).

Schiff (81, 82) and Lobene (83) reported the clinical effects of a dentifrice containing 3.3% soluble pyrophosphate derived from tetrasodium pyrophosphate in combination with 0.243% sodium fluoride and a silica abrasive versus a placebo dentifrice containing no pyrophosphate. Both trials were double-blind, randomized assessments of subjects who had documented histories of supragingival calculus accumulation, and both used the V.M.I. calculus assessment method. The investigations used those persons who were found to have the highest pretest calculus scores. For the

Table 7. Studies of Crystal Growth Inhibitors: Soluble Pyrophosphates

Study	Conc.	No.	Length	% Reduction
Kinoshita et al. (1966)[78]	2.9%	10	7 Days	15 (N.S.)
Zacherl et al. (1985)[79]	3.3%	418	6 Months	32
Mallatt et al. (1985)[80]	3.3%	217	2 Months	26
Schiff (1986,87)[81,82]	3.3%	122	6 Months	46
Lobene (1986)[83]	3.3%	120	3 Months	44
Grossman et al. (1987)[84]	3.3%	174	3 Months	Sign.
Lobene (1987)[85]	3.3%	107	3 Months	48
Juliano et al. (1986)[86]	3.3%	169	3 Months	40
Schiff (1987)[87]	3.3%	119	3 Months	29–49
Rosling & Lindhe (1987)[88]	3.3%	161	6 Months	9–42
Lobene (1987)[91]	3.3%	200	3 Months	51
Rustogi et al.(1987)[64]	3.3%	107	3 Months	32
Putt & Kleber (1988)[36]	3.3%	381	3 Months	5 (N.S.)
Kazmierczak et al.(1988)[65]	3.3%	197	3 Months	21
Rugg-Gunn et al.(1988)[92]	2.35%	32	6 Months	45–51
Rustogi et al.(1988)[93]	1.0%	93	3 Months	35
Singh (1988)[94]	1.0%	80	3 Months	38
Lu et al. (1988)[95]	5.0%	206	4 Months	15*

* as compared to 3.3% pyrophosphate

trial by Schiff, 122 subjects were utilized; for that by Lobene, 120 subjects were used. Subjects were randomly assigned to the test or control group, and received a dental prophylaxis. They were instructed to use their assigned dentifrice twice daily for one minute each time. At the end of three months, Schiff (81) and Lobene (83) reported 35.5% and 44.2% reductions in calculus deposits, respectively. After six months Schiff (82) reported 46.0% less calculus with the pyrophosphate dentifrice.

A double-blind, parallel group investigation by Grossman et al. (84) evaluated the ability of a dentifrice containing soluble pyrophosphate and fluoride to reduce calculus formation following the use of a 0.12% chlorhexidine mouthrinse. The panel included 174 adults who were instructed to rinse twice daily with chlorhexidine (15 ml. for 30 seconds each time) followed by toothbrushing and flossing in their usual manner. After three months the group using the soluble pyrophosphate dentifrice had a significant reduction in calculus occurrence.

Lobene (85) reported the results of a double-blind, randomized clinical trial comparing a dentifrice containing 3.3% soluble pyrophosphate and 0.243% sodium fluoride to a placebo dentifrice containing no pyrophosphate. The soluble pyrophosphate was derived from a mixture of 4.5% tetrapotassium pyrophosphate and 1.5% tetrasodium pyrophosphate. Known calculus formers were stratified on calculus scores (V.M.I.) and randomly assigned to one of the two test groups. At the end of three months, the results observed in the 107 panelists indicated a 47.6% reduction in calculus formation. Juliano et al. (86) found a 39.7% reduction in calculus in a three-month trial using an identical formulation in a panel of 169 subjects.

Two clinical trials reported by Schiff (87) and by Rosling and Lindhe (88) compared the efficacy of two dentifrices containing 3.3% soluble pyrophosphate provided from different soluble salts, one of which also contained 10% of a copolymer of methoxyethylene and maleic acid (Gantrez). This copolymer has been shown to be effective in inhibiting the seeded crystal growth of hydroxyapatite in vitro and in animal models (89, 90). Panelists were selected based on the amount of calculus present as determined from their pretest scores. In the study by Schiff, 119 subjects who were exceptionally heavy calculus formers, were selected while 161 subjects began the study by Rosling and Lindhe. In both studies the panelists were stratified by their calculus scores and assigned to one of three test groups. These groups used one of the two active products or a control dentifrice. All panelists were instructed to use their assigned dentifrice twice daily. The panelists were then re-examined after three and six months in the study by Rosling and Lindhe (88) and after three months in the Schiff study (87).

In the Schiff study (87) both dentifrices containing 3.3% soluble pyrophosphate with and without the copolymer were effective in reducing the prevalence of calculus deposits when compared to the placebo. Percent reductions were 49.3% and 29.4%, with the dentifrice containing the copolymer providing the greater reduction. After three months, Rosling and Lindhe (88) found no significant differences among the three groups in terms of calculus deposits; however, after six months the panelists using the active dentifrice with the copolymer had a significant reduction of 42.2% in calculus deposits as compared to a numerical reduction of 9.0% for the other active dentifrice.

In 1987 Lobene (91) published the results of a clinical study designed to compare the anticalculus effects of two dentifrices with a control. One active dentifrice contained 3.3% soluble pyrophosphate, 1% Gantrez, and 0.243% sodium fluoride in a silica base; the other active dentifrice contained 0.5% zinc citrate and 1.2% sodium monofluorophosphate. The control dentifrice contained 0.243% sodium fluoride. Panelists were instructed to use their assigned dentifrice twice daily for one minute each time, with three morning toothbrushings per week performed under supervision. Two-hundred subjects were then re-evaluated at the end of a three-month test period. The results indicated that the group using the dentifrice containing soluble pyrophosphate had a 50.8% reduction in calculus compared to the control. When compared to the zinc citrate dentifrice, the soluble pyrophosphate dentifrice resulted in 44.9% less calculus at the end of three months; the zinc citrate dentifrice produced no appreciable reduction in calculus deposits.

Rustogi et al. (64) compared a 3.3% soluble pyrophosphate dentifrice and a 2.0% zinc chloride formulation for calculus prevention; as compared to a placebo dentifrice, the soluble pyrophosphate dentifrice resulted in a significnat reduction in calculus of 32%. Putt and Kleber (36) included a 3.3% pyrophosphate dentifrice as a positive control in their study of aluminum citrate cited earlier; after three months they reported a numerical decrease in calculus formation of 4.5%, which was not statistically significant. Kazmierczak et al. (65) also evaluated a 3.3% soluble pyrophosphate dentifrice and observed a 21% decrease in calculus formation during a 3-month test period.

The use of a dentifrice containing 2.35% soluble pyrophosphate in combination with 0.8% sodium monofluorophosphate (NaMFP) has also been reported by Rugg-Gunn et al. (92) as being effective in the control of dental calculus in a six-month clinical trial involving a crossover design with 32 adults. The results indicated that the dentifrice containing pyrophosphate reduced the deposition of dental calculus by 45% with the Volpe-Manhold Index and by 51% using the Marginal Line Index.

Recent studies by Rustogi et al. (93) and Singh (94) have also shown the efficacy of a mouthrinse containing soluble pyrophosphate and a copolymer to control supragingival dental calculus. Both studies were three-month, double-blind investigations involving 93 and 80 participants, respectively, who were balanced on their baseline calculus scores (V.M.I.) and randomly assigned to use either the active or a control mouthrinse. In addition to their assigned mouthrinse, all panelists were instructed to use a dentifrice containing 0.76% Na_2PO_3F. The participants were instructed to brush twice daily for one minute followed by a one minute rinse using 10 milliliters of their assigned solution. The use of the mouthrinse containing the active agents resulted in reductions of 34.7% and 37.7% in the two studies, respectively.

A recent investigation by Lu et al.(95) compared the use of a

3.3% pyrophoshate dentifrice to a dentifrice containing 5% pyrophosphate. Both dentifrices contained 1100 ppm fluoride as NaF. The four-month trial involved 206 adults who were stratified by whole-mouth V.M.I. scores, sex, and age. The panelists were instructed to use their assigned dentifrice at least once per day in their normal manner. The four-month results indicated that those panelists using the 5% formulation developed 14.5% less calculus than those using the dentifrice containing 3.3% pyrophosphate. In addition, among heavy calculus formers (V.M.I.>12.0) the data indicated that the group using the 5% dentifrice developed 23.3% less calculus than those using the 3.3% formula.

Conclusions

A number of chemical agents have been evaluated to determine their ability to either prevent or suppress the clinical formation of dental calculus. While it is impossible for a number of reasons, to compare the various modalities directly the use of crystal growth inhibitors, notably soluble pyrophosphates, diphosphonates, and compounds containing zinc, offers the best means of controlling dental calculus at the present time. Of these agents, the use of diphosphonates and soluble pyrophosphate has been the most extensively studied and their efficacy for calculus prevention has been clearly documented.

Bibliography

1. Marshall-Day, C.D.; Stephens, R.G.; and Quigley, L.F. (1955) J. Periodontol. 26:185-203.
2. Field, H.M., Hawkins, B., Lainson, P.A., Townsend, B., Beck, J.D., Walker, J., Ettinger, R. (1982) J. Dent. Res. 61 (Sp.Iss.):179(Abst. No. 6).
3. Miller, A.J.; Brunelle, J.A.; Carlos, J.P.; Brown, L.J.; and Loe, H. (1987): Oral Health of United States Adults, The National Survey of Oral Health in U.S. Employed Adults and Seniors: 1985-1986, U.S. Department of Health and Human Services, Public Health Service, National Institutes of Health: NIH Publication No.87-2868.
4. Löe, H. (1983) Int. Dent. J. 33:119-126.
5. Stewart, G.G. (1952) J. Periodontol. 23:85-90.
6. Aleece, A.A. and Forscher, B.K. (1954) J. Periodontol. 25:122-125.
7. Jensen, A.L. (1959) JADA, 59:923-930.
8. Ennever, J. and Sturzenberger, O.P. (1961) J. Periodontol. 32:331-333.
9. Packman, D.E.; Abbott, D.D.; Salisbury, G.B.; and Harrisson, J.W.E. (1963) J. Periodontol. 34:255-258.
10. Harrison, J.W.E.; Salisbury, G.B.; Abbott, D.D.; and Packman, E.W. (1963) J. Periodontol. 34:334-337.
11. Kerr, D. and Field, H. (1944) J. Dent. Res., 23:313-316.

12. Warren, E.B.; Hansen, N.; Swartz, M.; and Phillips, R. (1964) J. Periodontol. 35:505-512.
13. Grossman, L.I. (1954) Oral Surg., Oral Med. and Oral Path. 7:607-608.
14. Gunson, J.J. (1955) Dent. Surv., 31:1248-1250.
15. Weinstein, E. and Mandel, I. (1964) J. Oral Ther. and Phar., 1:327-334.
16. Dossenbach, W.F. and Muhlemann, H.R. (1961) Helv. Odont. Acta., 5:25-28.
17. Muller, E.; Schroeder, H.E.; and Mühlemann, H.R. (1962) Helv. Odont. Acta., 6:42-45.
18. Kristoffersen, T. (1963) Odont. Tskr. 71:179-198.
19. Sastri, A. and Yuktanandana, I. (1966) Parodont. 20:28-35.
20. Westerholm, N. (1971) Suom Hammaslääk Toim. 67:149-151.
21. Volpe, A.R.; Kupczak, L.J.; Brant, J.H.; King, W.J.; Kestenbaum, R.C.; and Schlissel, H.J. (1969) J. Dent. Res. Suppl., 48:832-841.
22. Volpe, A.R.; Schulman, S.M.; Goldman, H.M.; King, W.J.; and Kupczak, L.J. (1970) J. Periodontol. 41:463-467.
23. Stallard, R.E.; Volpe, A.R.; Orban, J.E.; and King, W.J. (1969) J. Periodontol. 40:683-694.
24. Rokita, J.R.; Hazen, S.P.; Millen, D.; and Volpe, A.R. (1975) Phar. and Ther. in Dent. 2:1-11.
25. Cancro, L.P.; Paulovich, D.B.; Klein, K. and Picozzi, A. (1972) J. Periodontol. 43:687-691.
26. Löe, H.; Mandell, M.; Derry, A.; and Rindom Schiött, C. (1971) J. Periodontal Res., 6:312-314.
27. Löe, H., Rindom Schiött C.; Glavind, L; and Karring, T. (1976) J.Periodontal Res. 11:135-144.
28. Hirsh, S., and Leclercq, J. (1978) L'Information Dentaire, 60:21-31.
29. Albertini, Cup de Saint-Paul, Goldberg, Sanguy and Tyrand (1983) L'Information Dentaire, 65:853-855.
30. Belting, C.M. and Gordon, D.L. (1966) J. Periodontol. 37:20-25.
31. Belting, C.M. and Gordon, D.L. (1966) J. Periodontol. 37:26-33.
32. Shapiro, W.B.; Kaslick, R.S.; Chasens, A.I.; and Eisenberg, R. (1973) J.Periodontal 44:636-639.
33. Son, S. and Muhlemann, H.R. (1971) Helv. Odont. Acta. Suppl. VII, 15:158-159.
34. Mandel, I.M. and Thompson, R.H. (1967) J.Periodontal 38:310-315.
35. Swenson, H.M. and Bixler, D. (1967) J. Periodontol. 38:481-484.
36. Putt, M.S. and Kleber, C.J. (1988) J. Dent. Res. 67:246 (Abst. No. 1067).
37. Tsuprova, N.D. (1968) Stomatologiia, 47:83-84.
38. Turesky, S. ; Gilmore, N.D.; and Glickman, I. (1967) J. Periodontol. 38:142-147.
39. Gilmore, N.D.; Turesky, S.; and Glickman, I. (1968) J. Periodontol. 39:284-285.
40. Mühlemann, H.R.; Bowles, D.; Schiatt, A.; and Bernimoulin, J.P. (1970) Helv. Odont. Acta. 14:31-33.

41. Sturzenberger, O.P.; Swancar, J.R.; and Reiter, G. (1971) J. Periodontol. 42:416-419.
42. Conroy, C.W.; Sturzenberger, O.P.; Bollmer, B.; Swancar, J.R.; and Zimmerman, E.R. (1972) J. Dent. Res., (Abst. No. 208).
43. Suomi, J.D.; Horowitz, H.S.; Barbano, J.P.; Spolsky, V.W.; and Heifetz, S.B. (1974) J. Periodontol. 45:139-145.
44. Videmari, N. (1977) Rev. Assoc. Odont. Argent., 65:95-102.
45. Herforth, A.; Fligge, U.; and Strassburg, M. (1977) Dtsch. Zahnärtzl. Z., 32:757-759.
46. König, D. (1978) Dissertation, Universität Düsseldorf.
47. Laturnus, J. (1979) Dissertation, Universität Düsseldorf.
48. Seichter, U.; Chatzigiannis, S.; Herforth, A.; and Schübel, F. (1979) Prog. and Abstracts, European Division, Int. Assn. Dent. Res., (Abst. No. 87).
49. Kremers, L.; Daliemunthe, S.H.; and Lampert, F. (1980) Dtsch. Zahnärtzl. Z. 35:729-731.
50. Halbritter, P.F.M. (1973) Inaugural Dissertation, Universität Zürich.
51. Herforth, V.A. (1976) Disch. Zahnärztl. Z., 31:392-395.
52. Seichter, U., and Herforth, A. (1980) 27th ORCA Congress (Abst. No. 21).
53. Seichter, U. and Laturnus, J. (1980) Int. Assn. Dent. Res. (Abst. No. 65).
54. Cassese, G.; Celeste, G.; DeNotaris, V.; Gargiulo, V.; Serra, G.; and Silvano, G. (1980) Estratto da Medical Praxis, 1:221-227.
55. Bubani, G. (1980) Estratto da Medical Praxis, 1:215-220.
56. Nilsson, B., and Holm, G. (1986) Tandläkartidningen, 78:886-888.
57. Arends, J. and Dijkman, A.G. (1987) presented at the 34th ORCA Congress (Abst. No. 128).
58. Koch, G.; Karlsson, R.; Bergman-Arnadottir, I.; Bjarnason, S.; Finnbogason, S.; and Höskuldsson, O. (1988) presented at the 35th ORCA Congress (Abst. No. 110).
59. Hanke, M.T. (1940) JADA, 27:1379-1393.
60. Picozzi, A.; Fischman, S.L.; Pader, M.; and Cancro, L.P. (1972) J. Periodontol. 43:692-695.
61. Kohut, B. and Grossman, E. (1986) J. Dent. Res., 65:275 (Abst. No. 951).
62. Schmid, M.O.; Schiatt, A.; and Muhleman, H.R. (1974) Helv. Odont. Acta., 18:22-24.
63. Lobene, R.R.; Soparkar, P.M.; Newman, M.B.; and Kohut, B.E. (1987) J.A.D.A., 114:350-352.
64. Rustogi, K.N.; Volpe, A.R.; and Petrone, M.E. (1988) Compend. Cont. Dent. Ed., 9:78-79.
65. Kazmierczak, M.; Mather, M.L.; Ciancio, S.G.; Fischman, S.; and Cancro, L. (1988) J. Dent. Res., 67:246 (Abst. No. 1068).
66. Stephen, K.W.; Burchell, C.K.; Huntington, E.; Baker, A.G.; Russell, J.I.; and Creanor, S.L. (1987) Caries Res., 21:380-384.
67. Fleisch, H. and Bisaz, S. (1962) Nature, 195:911
68. Fleisch, H.; Russell, R.G.G.; Bisaz, S.; Termine, J.D.; and

Posner, A.S. (1968) Calcif. Tiss. Res., 2:49-59.

69. Fleisch, H. and Russell, R.G.G.(1972) J. Dent. Res. Suppl. to No. 2, 51:324-332.
70. Vogel, J.J. and Amdur, B.H. (1967) Arch. Oral Biol., 12:159-163.
71. Edgar, W.M. and Jenkins, G.N. (1972) Arch. Oral Biol., 17:219-223.
72. Mukherjee S. (1968) J. Perodontal Res. Suppl. No. 2, 3:13-35.
73. Briner, W.W. and Francis, M.D. (1973) Calif. Tiss. Res., 11:10-22.
74. Lu, K.H.; Yen, D.J.C.; Zacherl, W.A.; Ruhlman, C.D.; Sturzenberger, O.P. and Lehnhoff, R.W. (1985) J. Dent. Child. 52:449-451.
75. Strang, R.; Douglas, E.; Damato, F.A.; and Stephen, K.W. (1988) presented at the 35th ORCA Congress (Abst. No. P112).
76. White, D.J. and Faller, R.V. (1986) Caries Res., 10:332-336.
77. Mellberg, J.R.; Castrovince, L.A.; Rotsides, I.D.; Hayes, J.C. and Deutchman, M. (1987) Compend. Cont. Dent. Educ. Suppl. No. 8, 8:s267-271.
78. Kinoshita, S. and Muhlemann, H.R. (1966) Helv. Odont. Acta., 10:46-48.
79. Zacherl, W.A.; Pfeiffer, H.J.; and Swancar, J.R. (1985) J.A.D.A., 110:737-738.
80. Mallatt, M.E.; Beiswanger, B.B.; Stookey, G.K.; Swancar, J.R.; and Hennon, D.K. (1985) J. Dent. Res. 64:1159-1162.
81. Schiff, T.G. (1986) Clin. Prev. Dent., 8:8-10.
82. Schiff, T.G. (1987) Clin. Prev. Dent. 9:13-16.
83. Lobene, R.R. (1986) Clin. Prev. Dent., 8:5-7.
84. Grossman, E.; Sturzenberger, O.P.; Bollmer, B.W.; Moore, D.J.; Manhart, M.D.; and Huetter, T.E. (1987) J. Dent. Res., 66:279 (Abst. No. 1383).
85. Lobene, R.R. (1987) Compend. Cont. Dent. Educ., 8:175-178.
86. Juliano, G.F.; Villanueva, E.H.; Carlos, V.M. (1986) J. Philip. Dent. Assoc. 36:28-30.
87. Schiff, T.G. (1987) Compend. Cont. Dent. Educ. Suppl. No. 8, 8:s275-277.
88. Rosling, B. and Lindhe, J. (1987) Compend. Cont. Dent. Educ. Suppl. No. 8, 8:s278-282.
89. Gaffar, A. and Esposito, A. (1986) J. Dent. Res. 65:771 (Abst. No. 409).
90. Gaffar, A.; Schmid, R. Afflitto, J.; and Coleman, E. (1987) Compend. Cont. Dent. Educ. Suppl. No. 8, 8:s251-255.
91. Lobene, R.R. (1987) Clin. Prev. Dent., 9:3-8.
92. Rugg-Gunn, A.J. (1988) Br. Dent. J. 165:133-136.
93. Rustogi, K.N.; Volpe, A.R. and Petrone, M.E. (1988) J. Dent. Res., 67:246 (Abst. No. 1071).
94. Singh, S.M. (1988) Am.J.Dent. 1:9-11.
95. Lu, K.H.; Ruhlman, C.D.; Chung, K.; Adams, A.; and Bollmer, B.W. (1988) J. Ind. Dent. Assoc., 67:17-18.

G.Koch[1]
I.Bergmann-Arnadottir[2]
S.Bjarnason[3]
S.Finnbogason[4]
O.Höskuldsson[2]
R.Karlsson[1]

A three-year controlled clinical trial on caries-preventing effect of fluoride dentifrices with and without anticalculus agents

[1]The Institute for Postgraduate Dental Education, Jönköping, Sweden, [2]Faculty of Odontology, University of Iceland, Reykjavik, Iceland, [3]Faculty of Odontology University of Gothenburg, Sweden and [4]School Dental Services, Reykjavik, Iceland

ABSTRACT

A three year, double blind, randomized caries trial was conducted to evaluate the relative anticaries efficacy of four different sodium fluoride dentifrices containing 250 ppm fluoride, 1000 ppm fluoride, 1000 ppm fluoride in combination with 1% disodium 1-hydroxyethylidene-1.1-bisphosphonate and 1000 ppm fluoride in combination with 1% disodium azacycloheptylidene-2.2-bisphosphonate. As a positive control a monofluorophosphate dentifrice (1000 ppm fluoride) was used.

1161 Icelandic children, 11 and 12 years of age, were randomly assigned to one of the five treatment groups. 1035 subjects completed the trial. After three years of unsupervised brushing, the dentifrice containing 250 ppm fluoride was significantly less effective in controlling caries increment. The addition of the anticalculus agents had no detrimental influence on the caries preventing effect of the sodium fluoride dentifrice. In contrast the combination of sodium fluoride and disodium azacycloheptylidene-2.2-bisphosphonate was significantly more effective than the positive control.

Recent Advances in the Study of Dental Calculus

INTRODUCTION
The caries inhibiting effect of standard fluoride dentifrices containing 1000 ppm fluoride is well documented (1-3). However, in an attempt to enhance cariostatic effectiveness, dentifrices containing higher fluoride concentrations or various combined fluoride formulations are being introduced. At the same time the growing exposure to fluorides has raised concerns about potential toxicological side effects (4). No increase in the prevalence of fluorosis related to dentifrice ingestion has been reported (5-7). For safety reasons, dentifrices with lower fluoride content have been introduced for use by preschool children in Scandinavia and several other countries. However, data documenting caries inhibiting efficacy of dentifrices containing less than 1000 ppm fluoride are sparse and contradictory (8-13).

Even more ambiguous regarding caries development have been the addition of anticalculus agents like bisphosphonates or pyrophosphate to dentifrices. These compounds, known as powerful inhibitors of crystal growth, reduce the rate of enamel dissolution, thus becoming potentially anticariogenic (14). Contradictory findings concerning bisphosphonates from in vivo and in vitro studies have been reported. The inhibited remineralization of enamel has been observed in vitro (15-17). No such phenomenon has been demonstrated when bisphosphonate effect on enamel in vivo was studied. The bisphosphonate HEBP reduced the mineral loss in enamel during the Intra Oral Cariogenicity Test and the effect was additive when fluoride and bisphosphonate were applied simultaneously (17,18). Reduction of caries and enhanced fluoride retention in enamel by bisphosphonate have been demonstrated in experimental animals (19). However, no long term clinical studies have been performed to test the effect of a dentifrice containing fluoride and bisphosphonate on caries.

The aim of the present clinical trial was to evaluate the relative anticaries efficacy of a dentifrice with reduced sodium fluoride content (250 ppm F), of a standard sodium fluoride dentifrice (1000 ppm F) and of two standard sodium fluoride dentifrices containing the phosphonates disodium 1-hydroxy-ethylidene-1.1-bisphosphonate (HEBP) or disodium azacyclo-heptylidene-2.2-bisphosphonate (AHBP), respectively, against a standard sodium monofluorophosphate dentifrice (1000 ppm F) as positive control.

MATERIAL AND METHODS

Study Site and Population
Reykjavik, Iceland was the chosen study site, as caries prevalence in Iceland is among the highest reported internationally, and no decrease has been observed in the last decades despite the availability of free dental care to all children of school age (20). Fluoride concentration in the public water supply of Reykjavik is below 0.1 ppm, and no regular preventive fluoride programs have been established in schools.

Table I. Treatment groups and dentifrices

Group	Fluoride source	Abrasive system	Fluoride, ppm start	3 yrs.	Anticalc. agent	pH
I[a]	NaF	silica	250	250	none	7.0
II[b]	MFP	$CaHPO_4 \cdot 2H_2O$	940	700	none	7.5
III[c]	NaF	silica	980	980	AHBP	7.0
IV[a]	NaF	silica	970	950	none	7.0
V[d]	NaF	silica	940	930	HEBP	6.8

[a] Experimental dentifrice, Henkel KGaA, Düsseldorf, FRG
[b] Colgate Great Regular Flavor, Colgate Palmolive Company, New York, USA
[c] Thera-med II, current market product, Henkel KGaA, Düsseldorf, FRG
[d] Thera-med I, former market product, Henkel KGaA, Düsseldorf, FRG

Study Design

Prior to the study, the decision was made that the differences between the treatment groups in caries preventive effect should be about 15 per cent in mean total increment of decayed and filled tooth surfaces to be of clinical significance. The determination of sample size was based on pre-experimental analysis of caries epidemiological data compiled by the School Dental Service of Reykjavik, showing an average annual caries increment among 11-14 year-old children of about 4 decayed and filled surfaces. The adequate sample size was calculated to be at least 200 individuals in each treatment group in order to ensure 80% power with the level of significance (21) set at 0.05 accounting for a drop out rate of 10% during the three year study period. The trial was unsupervised, double blind and conducted according to the principle requirements outlined by the Federation Dentaire Internationale and the Declaration of Helsinki (22).

Test Subjects

In 1983, 1161 children 11 and 12 years of age, started the study. The participants were stratified according to age and school and then randomly assigned to one of the five treatment groups. Siblings were allocated to the same group.

Dentifrices

Table I presents the dentifrices used in the study. A fluoride dentifrice with well documented anticaries effect was used as a positive control. A batch of each of the four test dentifrices was produced with an appearance and flavor similar to the positive control dentifrice, which was purchased and subsequently refilled in 75 ml laminated tubes, identical for all dentifrices in this study. Sufficient amounts of the assigned dentifrice to use exclusively during the study, were delivered to the participants and their families twice a year by appointed distributors.

Clinical Examinations

The examinations were performed by two examiners. Calibration sessions were conducted prior to baseline and at each of the annual examinations in order to standardize clinical and radiographic diagnosis and recording procedures. All study participants were randomly allocated to the examiners, who examined the same children at the baseline and thereafter annually for three years. All examinations were conducted in dental clinics using a standard operatory light. The clinical examinations were supplemented by posterior bitewing radiographs taken and processed under standardized conditions.

Diagnostic Criteria

A tooth was designated erupted when any part of its crown had penetrated the gingiva. All accessible permanent tooth surfaces were examined clinically. Proximal tooth surfaces from the distal surface of the canine to the mesial surface of the second molar in each quadrant were diagnosed on the bitewing radiographs. The following criteria were used:
Initial clinical caries: White demineralized areas without enamel breakdown on an unrestored tooth surface.
Manifest clinical caries: Pits and fissures, catching the probe and cavities on an unrestored tooth surface.
Initial radiographic caries: Clearly defined radiolucency in the outer half of the enamel on an unrestored tooth surface.
Manifest radiographic caries: Radiolucency exceeding half of the enamel thickness on an unrestored tooth surface.
Restored tooth surface: Any restoration of tooth surfaces.
A carious or filled pit or fissure on the buccal surface of the lower molars and/or on the lingual surface of the upper molars was recorded as occlusal.

Statistical Analysis

Only new manifest lesions and fillings were employed for the evaluation of the caries preventive effect of the dentifrices. Teeth extracted during the experimental period were scored according to their status at the last observation. Net caries increment was expressed as decayed and filled surfaces minus reversals.

A one-way analysis of variance (ANOVA) was performed to test group comparability at baseline examination concerning caries prevalence, surfaces at risk or gingivitis as well as to test the significance of differences of new decayed and filled tooth surfaces among the five dentifrice groups. If the ANOVA rejected the multisample null hypothesis of equal means, multiple comparison testing using Duncan's test was performed to explore where the differences were located. A multivariate analysis of variance (MANOVA) was also used. Categorical variables such as

Table II. Number of children at randomization and at each of the four examinations

Checkpoint	Group					Total	Drop outs
	I	II	III	IV	V		
Randomization	231	232	237	231	230	1161	
Baseline	229	229	234	229	225	1146	15
1 year	214	221	226	217	212	1090	56
2 year	206	216	220	211	210	1063	27
3 year	203	209	211	209	203	1035	28

Table III. Results of the examiner reliability tests

Examination	Difference between examiners (p)	Intraclass correlation coefficient (ICC)	Lower limit of 95% confidence interval of ICC
Baseline	> 0.05	0.992	0.980
1 year	> 0.05	0.996	0.990
2 year	> 0.05	0.998	0.996
3 year	> 0.05	0.993	0.990

sex and age were tested with chi-square test. All statistical tests were two-tailed and at the 5 percent significance level.

RESULTS
The number of children and their distribution to the test groups at randomization and at each examination is presented in Table II.
A total of 1035 subjects completed the study, whereby each group contained more than 200 individuals. There were no statistically significant differences between the groups regarding number of children, sex and age distribution, surfaces at risk or decayed and filled tooth surfaces at baseline. There were also no statistically significant differences concerning number of children and the distribution of sex and age at any of the examinations. 126 children (10.9%) withdrew between randomization and the three-year examination.
The following data and calculations are based on children who completed the three year study. Due to the high consistency between the examiners (Table III), their data were pooled for statistical analysis.

Table IV. Three year net increment [Mean (± SD)] of decayed and filled tooth surfaces (DFS) and of tooth surfaces with initial caries

Subjects	Group					Differences
	I	II	III	IV	V	
Manifest						
Total[a]	12.7	10.9	8.8	10.1	10.1	I > II,III,IV,V
	(9.3)	(7.9)	(7.2)	(8.1)	(8.4)	II > III
Boys[b]	13.0	10.7	8.6	10.0	10.2	I > II,III,IV,V
	(10.3)	(8.2)	(6.6)	(7.1)	(8.6)	
Girls[c]	12.3	11.0	9.0	10.1	10.0	I > III
	(8.2)	(7.6)	(7.7)	(9.0)	(8.2)	
Born'71[b]	13.8	11.1	9.0	11.4	10.9	I > II,III,V
	(10.7)	(7.8)	(6.6)	(8.6)	(9.1)	
Born'72[c]	11.7	10.7	8.6	9.0	9.4	I > III,IV,V
	(7.9)	(8.1)	(7.7)	(7.4)	(7.7)	
Initial						
Total	7.7	8.4	7.0	7.2	5.8	none
	(10.4)	(9.6)	(9.4)	(9.8)	(9.6)	

ANOVA: [a] $p<0.001$, [b] $p<0.01$, [c] $p<0.05$

Variable	MANOVA (p)	Differences
Dentifrice	< 0.001	I > II,V,IV > III
Year	< 0.001	3 > 2 > 1
Age	< 0.01	71 > 72
Sex	> 0.05	none
Examiner	> 0.05	none

Table V. Multiple analysis of variance for the net increment of decayed and filled surfaces (DFS)

Caries Increment

Group I showed a significantly higher caries increment (DFS) compared to all other groups. The children in Group II (positive control) developed 14.2 % and those in Group III 30.7 % less caries, compared to Group I.

A significant difference was also found between groups II and III. In comparison the caries increment for children in Group III was 19.3 % lower. In general Group III showed the lowest caries increment in total material, as well as in age and sex subgroups. No differences between the treatment groups were observed regarding the increment of initial carious lesions (Table IV). Employing MANOVA, with dentifrice, year, age, sex and examiner as variables, the annual caries increment for group I was significantly higher and that for group III significantly lower compared with other groups.

In addition significant differences were found with respect to year of examination and age of children. No differences were seen concerning sex or examiner (Table V).

DISCUSSION

Due to ethical considerations no fluoride-free dentifrice could be tested. A monofluorophosphate dentifrice, recognized by the Council on Dental Therapeutics of the American Dental Association as a clinically effective fluoride dentifrice, was used as a positive control. Thus only relative anti-caries efficacy of the test dentifrices could be established. The relatively large caries increment, examiner diagnostic consistency, and low attrition over the trial period ensured sufficient statistical power to determine clinically important differences between the test and positive control dentifrices.

The results of this study indicate that a sodium fluoride dentifrice containing 250 ppm fluoride is significantly less effective in preventing caries development than an identical formulation containing 1000 ppm. These findings are in agreement with results reported in a recent study, where anticaries efficacy of 250 and 1000 ppm fluoride as monofluorophosphate were compared (12). A similar dose-response relationship was reported by Reed (8) and in vitro as well as in vivo tests have shown favorable results regarding fluoride uptake and enamel de- and remineralization for the higher fluoride concentrations (18,23,24).

No significant differences in caries reduction between dentifrices with low and high fluoride content have been found by Forsman (9), Gerdin (10), Koch et al. (11) and Petersson et al. (13). Contrary to the neutral pH of the dentifrices employed in this study, Gerdin (10), Koch et al. (11) and Petersson et al. (13) used formulations with a pH of 5.5. Topical fluoride vehicles with lower pH have been found to promote the formation of calcium fluoride on the enamel surface, and presumably enhance enamel remineralization by prolonged release of fluoride (25). Therefore the low pH of the latter dentifrices might explain the differences in clinical effect.

Several clinical studies have indicated that sodium fluoride might have some advantages in caries prevention compared to monofluorophosphate (26,27). With 7% difference between Group II and IV, a similar tendency was observed in this study. The decrease in available fluoride in the MFP-dentifrice during the three year storage time may have contributed to this result (Table I).

The addition of the anticalculus agents HEBP and AHBP had no detrimental influence upon the caries preventive effect of the sodium fluoride dentifrice. In connection with the question, if anticalculus agents would interfere with remineralization it is of particular interest that no increase in initial carious lesions could be detected. These findings are in agreement with former conclusions drawn by Plöger and Klüppel (14) and with in vivo studies where no disturbances in remineralization by bisphosphonates were observed in the presence of the salivary pellicle (17,28). The combination of AHBP and fluoride proved superior to all other tested formulations in reducing caries increment, corroborating earlier observations that fluoride and bisphosphonate act synergistically (26,5).

The possible coating of calcium fluoride with bisphosphonate may provide a pH dependent slow release system that might enhance the cariostatic properties of topically applied fluoride as it has been suggested for pyrophosphate in a recent study (29). The differences between HEBP and AHBP may depend on different solubility of the calcium bisphosphonate layers formed onto the enamel surface. Additionally, due to the different chemical structure of the compounds, the surface layers may alter the surface charges and thus influence ion exchange to a varying degree.

REFERENCES
1. DePaola, P.F. (1983) Caries Res. 17, Suppl. 1, 119-135.
2. Stookey, G.K. (1985) in Clinical uses of fluorides, (ed. Wei) pp 105-131. Lea & Febiger, Philadelphia, USA.
3. Glass, R.L. (1986) J. R. Soc. Med. 79, Suppl 14, 15-17.
4. Whitford, G.M., Allman, D.W. and Shahed, A.R. (1987) J. Dent. Res. 66, 1072-1078.
5. Holm, A.K. and Andersson, R. (1982) Community Dent. Oral Epidemiol. 10, 335-339.
6. Driscoll, W.S., Horowitz, H.S., Meyers, R.J., Heifetz, S.B., Kingman, A. and Zimmermann, E.R. (1986) J. Amer. Dent. Ass. 113, 29-33.
7. Soparkar, P.M. and DePaola, P.F. (1985) J. Dent. Res. 64, 226 (Abstract 459).
8. Reed, M.W. (1973) J. Amer. Dent. Ass. 87, 1401-1403.
9. Forsman, B. (1974) Community Dent. Oral Epidemiol. 2, 166-175.
10. Gerdin, P.O. (1974) Swed. Dent. J. 67, 283-297.
11. Koch, G., Petersson, L.G., Kling, E. and Kling, L. (1982) Swed. Dent. J. 6, 233-238.
12. Mitropoulos, C.M., Holloway, P.J., Davies, T.G.H. and Worthington, H.V. Community Dental Health 1984, 1, 193-200.
13. Petersson, L.G., Johansson, M., Jönsson, G., Birhed, D. and Gleerup, A. (1988) 35th Annual ORCA Congress, Angers (France), Abstract 109.
14. Plöger, W. and Klüppel, H.J. (1982) Kariesprophylaxe 4, 129-137.
15. Mühlemann, H.R. and Aeschbacher, M. (1970) Helv. Odontol. Acta 14, 30-31.
16 ten Cate, J.M., Jongbloed, W.L. and Arends, J. (1981) Caries Res. 15, 60-69.
17. Wöltgens, J.H.M. and Koulourides, T. (1983) Caries Res. 17, 357-364.
18. van Croonenburg, E.J., Wöltgens, J.H.M., Qua, C.J. and de Blieck-Hogervorst, J.M.A. (1986) J. Biol. Buccale 14, 231-234.
19. Regolati, B. and Mühlemann, H.R. (1970) Helv. Odontol. Acta 14, 37-42.
20. Möller, P. (1985) Community Dent. Oral Epidemiol. 13, 230-234.
21. Kingman, A. (1977) Community Dent. Oral Epidemiol. 6, 30-33.
22. Commission on Oral Health, Research and Epidemiology,

FDI Technical Report No.1 (1982) Int. Dent. J. <u>32</u>, 292-310.
23. Barbakov, F., Sener, B. and Saltini, C. (1986) Helv. Odontol. Acta <u>2</u>, 1504-1511.
24. De Kloet, H.J., Extercate, R.A.M., Rempt, H.E. and Ten Cate, J.M. (1986) J. Dent. Res. <u>65</u>, 1410-1414.
25. Rölla, G. (1987) Personal communication.
26. Edlund, K. and Koch, G. (1977) Scand. J. Dent. Res. <u>85</u>, 41-45.
27. Lu, K.H., Ruhlmann, C.D., Chung, K.L., Sturzenberger, O.P. and Lehnhoff, R.W. (1987) J. Dent. Child, July-Aug., 241-244.
28. Wöltgens, J.H.M., Qua, C. and Blieck-Hogervorst, J.M.A. (1984) J. Biol. Buccale <u>12</u>, 339-346.
29. Lagerlöf, F., Soregaard, E., Barkvoll, P. and Rölla, G. (1988) J. Dent. Res. <u>67</u>, 447-449.